Standard Methods of

CLINICAL CHEMISTRY

VOLUME II

Standard Methods of
CLINICAL
CHEMISTRY

VOLUME II

BY THE AMERICAN ASSOCIATION
OF CLINICAL CHEMISTS

Editor-in-Chief:
DAVID SELIGSON
University of Pennsylvania
Philadelphia, Pennsylvania

1958
ACADEMIC PRESS INC., *Publishers*
NEW YORK, N. Y.

ACADEMIC PRESS, INC.
111 Fifth Avenue, New York, New York 10003

United Kingdom Edition published by
ACADEMIC PRESS, INC. (LONDON) LTD.
Berkeley Square House, London W.1

LIBRARY OF CONGRESS CATALOG CARD NUMBER: 53-7099

Third Printing, 1969

PRINTED IN THE UNITED STATES OF AMERICA

DEDICATION

THIS VOLUME is dedicated by Clinical Chemists to those physicians who contribute to the progress of modern medicine by skillful application of chemical data.

PREFACE

This series of "Standard Methods of Clinical Chemistry" was successfully started with the publication of Volume I in 1953 under the editorship of Dr. Miriam Reiner. The policies established, as listed in the Preface and Foreword of Volume I, have been continued. The editorship for Volume II was given to Dr. Nelson Young who unfortunately had to discontinue his work for personal reasons. The executive committee of The American Association of Clinical Chemists headed by Colonel Monroe Freeman nominated the present editor to continue this series of "Standard Methods."

The clinical chemist usually works in a hospital milieu. His interests are service, research, and teaching. When these are conscientiously pursued, the quality of medical care and benefits to the medical staff and patients improve. The range of knowledge and skills required in the practice of clinical chemistry is wide and often the busy or less experienced analyst does not have the time to investigate new or difficult methods which he desires to use. This volume is designed to provide accurate and workable methods upon which the clinical chemist can rely. Although some of these methods may not be used for daily work they should be helpful as references for evaluating those in use. This volume is directed primarily to clinical chemists. However, we believe that pathologists, medical technicians, clinical investigators, chemists in other fields, students and others will find in this volume, useful methods, ideas, facts, and references.

The techniques of blood pH determination and flame photometry are the only ones that have not been specifically verified by referees. Their general nature and the successful use of closely similar techniques in the editor's laboratory gave assurance of their dependability.

In the choice of methods, accuracy was a guiding principle. Some are very easy to perform. Others are time-consuming but essential for the practice of medicine. Some are more difficult than methods currently in use but have the advantage of being accurate and useful as reference methods. Alternative procedures to those found in Volume I for calcium, chloride, and cholesterol are included in this volume because they involve different principles. The chapter on sodium and potassium presents a different approach. That on the phosphatases uses a different substrate and the conditions differ. The lipase method was chosen to supersede that in Volume I because it measures more specifically the lipase released into blood by an inflamed or obstructed pancreas than does the tributyrinase method of Volume I. Methods such as those for phosphatases or phosphatides use or

vii

refer to the phosphate method of Volume I while providing alternatives. The other methods are new to this series.

The hemoglobin and nitrogen chapters provide methods for making secondary, but accurate, standards for hemoglobin and proteins. Whereas the biuret methods for serum proteins are preferred to the Kjeldahl method for daily use, the latter is essential for establishing standards for the former. The Kjeldahl method is so basic to clinical chemistry that several adaptations, macro and micro have been included. In contrast to the methods which can be accurately standardized is the empirical cephalin-cholesterol flocculation test. The best that can be offered is a clear description of the reagents and its performance so that it can be carried out uniformly.

Several methods are acknowledged to be more time consuming than some in common use. However, they were chosen because they provide dependable performance. An example is the method for 17-hydroxycorticosteroids in plasma. Several of our methods offer alternatives or short cuts with adequate warnings. References to methods which the analyst might find useful because of his special problems are included as often as possible.

The quality of reagents used, purification, and primary and secondary standards have been described. When doubt exists reagent grade chemicals should be used. "Water" refers to distilled water. We have tried to use significant figures to indicate the range of accuracy required for reagents or measurements. Where information was available, stability of reagents is stated. Information on collection of specimens and storage is provided, wherever it is essential or available. Notes have been freely interspersed in the text to allow the contributor or referee to make clear his directions, specifications or point of view. References made to commercial sources of equipment or materials do not imply that competitive products were less adequate.

Expressions for indicating precision, accuracy and variability will become more uniform with subsequent volumes. A chapter on statistics and control will appear in Volume III and will be helpful as a guide for future contributions.

Discussion of clinical or physiological problems related to the subject of the various chapters has been kept to a minimum. Often pertinent references to clinical and physiological reviews have been included. Ranges of values in healthy persons and often in disease, as established by these methods, have been included. We hope that analysts will report their data to us so that we can accumulate these for statistical studies.

The royalties from this volume are assigned to The American Association of Clinical Chemists which in turn has defrayed some of the costs of editing. The Graduate Hospital of the University of Pennsylvania has graciously sustained the remainder. The editor is greatly indebted to Dr. Reiner for her efforts in making available her experience and records from Volume I. He is grateful to Dr. Young for his initial efforts which were helpful. The editor is indebted to Dr. John Reinhold for his advice and constant "ear." He is grateful to the members of the editorial committee who were helpful in many ways and who made special efforts to attend meetings in order to establish policy and select methods, authors, and referees. He is grateful to Academic Press for the help with this manuscript. The editor owes his thanks to Miss Jean Marino who helped with the detail work, and to Mrs. Ruth Dunn Brodsky and Mrs. Mary Kenney who handled the correspondence and filing. He is, above all, grateful to the contributors and checkers who made this Volume.

To those clinical chemists who study this series, contribute to its improvement, and regard these methods as stepping stones to better ones, we offer our thanks.

Philadelphia, Pennsylvania　　　　　　DAVID SELIGSON, *Editor*

January 18, 1958　　　　　　　　　　*Editorial Committee*
　　　　　　　　　　　　　　　　　　JOSEPH GAST
　　　　　　　　　　　　　　　　　　MARGARET KASER
　　　　　　　　　　　　　　　　　　MIRIAM REINER
　　　　　　　　　　　　　　　　　　JOHN REINHOLD
　　　　　　　　　　　　　　　　　　JOSEPH ROUTH

CONTENTS

CALCIUM (COMPLEXIMETRIC)*

Submitted by: FRANK W. FALES, Department of Biochemistry, Emory University, Emory University, Georgia

Checked by: PEACE PAUBIONSKY, Biochemistry Laboratory, Abington Memorial Hospital, Abington, Pennsylvania

DONALD G. REMP, Biochemistry Laboratory, Henry Ford Hospital, Detroit, Michigan.

Introduction

The direct compleximetric titration procedure for the determination of serum calcium has three distinct advantages over the isolation procedures: The results may be obtained in a fraction of the time needed for the isolation procedures; less serum is required; and the elimination of the precipitation and washing steps removes possible sources of error.

The groundwork for the compleximetric procedure was laid by Schwarzenbach *et al.* (1, 2), who investigated the chelation of divalent and trivalent cations by ethylenediaminetetraacetate (EDTA). The efficacy of using murexide (ammonium purpurate) as the indicator for compleximetric titrations was also pointed out by these authors. In 1951 and 1952, a number of investigators independently developed compleximetric procedures for the determination of divalent cations of serum by modifications of the procedures developed by Schwarzenbach *et al.* for determining water hardness. These investigators included Sobel and Hanok (3), Buckley *et al.* (4), and Elliot (5), of the United States; Holtz (6, 7) of the Netherlands; and Flaschka and Holasek (8), of Germany. In general, these proposed procedures utilized EDTA as the titrant and either murexide or Eriochrome Black T as the indicator. The usefulness of the latter dye for the determination of serum calcium is limited by the fact that it forms a colored complex more readily with magnesium than with calcium and the end point indicates the total of these two cations when serum is titrated directly. On the

---

* Based on the method of Fales (9).

1

other hand, murexide has a higher affinity for calcium than for magnesium (2) and at an alkaline pH it is a specific indicator for calcium when serum is titrated directly (9). Unfortunately, the end point of the titration is not sharp, but there is a gradual change from a red to a purple color. Also, when serum with a high icteric index is encountered, the end point is further obscured. Independently, several investigators developed spectrophotometric methods for determining the end point. These investigators included Fales (9), Kibrick et al. (10), and Dreven et al. (11). The method described here is a modification of one of these procedures. The end point can be determined with considerable precision and no difficulty is encountered with serums having high bilirubin content.

Principles

The principles involved in the procedure are as follows: (a) At a high alkaline pH, EDTA has a much greater affinity for calcium than for magnesium. Therefore all the calcium is chelated before any of the magnesium. (b) Cations for which EDTA has a greater affinity are either absent or are present in insignificant quantities. (c) The murexide has a strong affinity for calcium, forming with it a red calcium purpurate complex. (d) As the EDTA removes the calcium, there is a gradual shift from red to purple as the ratio of purpurate ions to calcium purpurate increases. The titration is best followed spectrophotometrically by using light in the red region of the spectrum (620 mμ) so that the absorbance is due almost exclusively to the purple component. At shorter wavelengths, the absorbance is the resultant of the light absorption of both components, and the titration curve becomes complex and is not easily interpreted. (e) Although a portion of the calcium in serum is coupled with protein and, in the alkaline medium, with phosphate, conditions are maintained such that the bound calcium is in dynamic equilibrium with the calcium ions in solution. Therefore, the total calcium of the serum is measured, since EDTA has a greater affinity for calcium than has protein or phosphate.

Reagents

1. *Sodium hydroxide solution, 0.05 M.* Dissolve 4.0 g. of reagent grade NaOH in salt-free water and bring to a final volume of 2 l.

2. Sodium chloride solution, 1.4 M. Dissolve 81.8 g. of reagent grade NaCl in water and bring to a volume of 1 l.

3. Sodium chloride solution, 0.14 M. Add 100 ml. of 1.4 M saline to a 1-l. volumetric flask and bring to volume with water.

4. Stock EDTA solution, 0.005 M. Add 1.85 g. of reagent grade disodium dihydrogen ethylenediaminetetraacetate and 100 ml. of 1.4 M NaCl to a 1-l. volumetric flask. Bring to volume with water.

5. Stock indicator suspension. Place 0.25 g. of murexide (ammonium purpurate) in an amber dropping bottle, add 5.0 ml. of water and 25.0 ml. of 95% ethyl alcohol. The stock indicator will keep for several months if stored in the refrigerator. The incorporation of alcohol into the reagent was suggested by the work of Williams and Moser (12).

6. Stock calcium standard, 1.000 g. calcium/100 ml. Place 2.497 g. of oven-dried reagent grade $CaCO_3$ into an evaporating dish. Carefully dissolve the carbonate in a minimal amount of 6 N HCl and evaporate to dryness on a steam bath. Dissolve the resulting $CaCl_2$ in water, wash into a 100-ml. volumetric flask, and make to volume.

7. Working calcium standard, 10.00 mg./100 ml. or 5.00 meq./l. To a 1-l. volumetric flask, add 10.00 ml. of stock calcium standard and 100 ml. of 1.4 M NaCl. Bring to volume with water.

Procedure

METHOD I

This method is adaptable for colorimeter cuvettes 15–19 mm. in outside diameter and having a capacity of at least 15 ml.

With volumetric pipets, place 0.5 ml. of 0.14 M NaCl in a cuvette which will serve as the control, 0.50 ml. of calcium working standard in a second cuvette, and 0.50 ml. of fresh serum obtained from non-hemolyzed blood in other cuvettes. Up to 5 serum samples may be titrated simultaneously. The alkaline indicator solution is then prepared. Measure out a sufficient quantity of 0.05 M NaOH solution to allow 15 ml. for each cuvette and add well-mixed stock indicator until the proper absorbance is obtained. Using 19-mm. cuvettes (15.5-mm. light path), add indicator until the solution gives a reading of from 18 to 25% transmittance (0.75 to 0.60 A.) as compared to a water reference at a wavelength of 620 mμ. With tubes of smaller diameter, the indicator solution should be regulated to a proportionately lower absorbance. No added sensitivity is gained by adding murexide in excess of the quantity that can combine with the free calcium in the solutions; in fact, the interpretation of the end point becomes more difficult under these conditions.[a] The titrant is then prepared by adding 18 volumes of alkaline indicator solution to 1 volume of 0.14 M NaCl solution and 1 volume of stock EDTA solution. Prepare sufficient titrant to allow 10 ml. for each cuvette. Extreme precision is not required in preparing the titrant since a standard is run with each set of determinations. Add 4.5 ml. of alkaline indicator solution to each cuvette followed by 1.00 ml. of titrant. After mixing, set the control at a convenient reading of the spectrophotometer (for example, 30% transmittance), using a wavelength of 620 mμ. Readings of the standard and of each test solution are then made, the instrument being balanced with the control as before between each reading to compensate for the fading of the indicator. Remove all the cuvettes from the instrument and add 1.00 ml. of titrant again to each cuvette including the control. Mix and make readings again after balancing the instrument with the control. Continue readings until the end point has been passed, after which no further decrease in transmittance occurs. Usually 7 or 8 additions of titrant are required.

If the readings are made on the per cent transmittance scale, use 1-cycle semilogarithmic paper. Plot the milliliters of titrant on the abscissa at intervals of at least 1 inch. The intercept of the steepest portion of the titration curve with the plateau is the end point (see Fig. 1.) Place a straight-edged ruler in position so that a line drawn with a sharp pencil held vertically passes directly through the two points (or more) that delineate the steepest (most nearly vertical) portion of the titration curve and extend the line until it intersects the plateau. If the calcium concentration is known to be high, several milliliters of titrant may be added before readings are started, since the steepest portion of the curve shifts to the right with increasing calcium concentrations. On the other hand, if the calcium concentration is very low, it is advisable to make a zero reading before adding any titrant. In practice, the need for doing this is easily

[a] If the alkaline indicator solution is of the proper concentration, close to the maximal difference between the absorbance of the control and the standard will be obtained and the steepest portion of the titration curve will occur somewhere between the second and fourth addition of titrant.

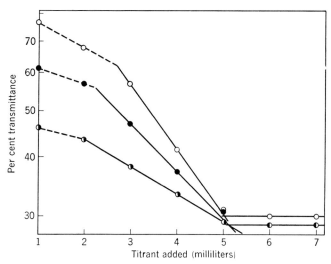

FIG. 1. Titration curves: ○, standard calcium solution; ●, serum alone; ◑ serum with added inorganic phosphate; and ◐, common points for serum alone and serum with added phosphate.

discerned, since the serum assumes a purple-red, rather than a red, color when the alkaline indicator solution is added.

METHOD II

This method is adaptable with cuvettes 18–24 mm. in outside diameter and having a capacity of at least 20 ml.

The titration is carried out exactly as in Method I, except that 1-ml. quantities of 0.14 M NaCl, standard, and serum are placed in the cuvettes and 9 ml. of alkaline indicator solution are added to each. In this case the titrant is prepared by diluting 1 volume of stock EDTA solution to 10 volumes with alkaline indicator solution. One-milliliter quantities of titrant are added as in Method I.

GENERAL CONSIDERATIONS

The titration may be carried out with a buret or a volumetric pipet for delivering the titrant. However, it is more convenient to use an automatic pipet. It is not necessary to regulate the pipet to deliver exactly 1-ml. quantities since a standard is run in conjunction with the serums, but the pipet must deliver the same volume with each ejection.

The following precautions should be taken. Once started, the titration must be carried to completion without delay since the calcium may slowly precipitate in the alkaline solution. Fresh serum must be used. When serum is stored, precipitation occurs and the bound calcium is no longer in mobile equilibrium with the free calcium. Serum from nonhemolyzed blood must be used. Apparently EDTA has a greater affinity for iron than has the porphyrin nucleus since high results are obtained with hemolyzed blood. Buckley *et al.* (4) added cyanide to counteract this, but this does not seem advisable to the author since, with gross hemolysis, the calcium level obtained would not be valid anyway (dilution effect). None of the cuvettes should remain in the light path for an extended period because the fading of the indicator is thereby enhanced.

Calculation

$$\text{Mg. \% Calcium} = \frac{\text{end point for serum}}{\text{end point for standard}} \times 10$$

$$\text{Meq./l. Calcium} = \frac{\text{end point for serum}}{\text{end point for standard}} \times 5$$

Discussion

The compleximetric method originally devised (9) utilized a highly alkaline medium which induced rapid fading of the murexide indicator. Also the color was diluted during the titration by the added standard EDTA solution. Under proper conditions, the control effectively compensated for these factors, but it was necessary to use a dilute indicator solution, or the fading and dilution masked the spectrophotometric measurement of the color change. The present method eliminates the dilution by employing a titrant containing the same indicator concentration as the test solution. Also the fading is minimized by utilizing a less alkaline medium. However, considerable time and effort were required to establish conditions whereby a less alkaline medium could be employed. Some of the findings in this study will be given as a warning and a guide for those who may be interested in devising a modification of the procedure.

It was found that the fading was reduced almost to the vanishing point when the alkalinity was reduced to 0.01 M NaOH (see also

Williams and Moser, 12) and that a standard calcium solution gave the identical end point as was obtained when the highly alkaline medium was employed (0.18 M NaOH). However, when the 0.01 M NaOH medium was employed for the titration of serum calcium, there was a marked shift of the apparent end point toward higher values. Since the author was confident that the original method measured the true calcium level, as evidenced by the results obtained during three years' continuous use in the clinical laboratory, it was necessary to conclude that some substance or substances in the serum interfered with the titration when the less alkaline medium was employed. When one considers the possible sources of interference, then the true complexity of this direct titration procedure is brought out. In the first place, the calcium exists both in the free ionic form and combined with protein. Upon bringing the solution to 0.01 M NaOH, the pH of the solution approaches the pK_3 of phosphoric acid, so that the inorganic phosphate present will be represented by about equimolar quantities of dibasic and tribasic ions. When the product of the calcium and tribasic phosphate ions is calculated, using normal values for the calcium and phosphate concentrations, a value of about 10^{-20} is obtained. This is far in excess of the solubility product of calcium phosphate. According to Logan and Taylor (13), the solubility product of calcium phosphate is $10^{-23.5}$ when no solid is present and 10^{-27} when $Ca_3(PO_4)_2$ crystals are present. However, no frank precipitation takes place. The following factors may be operative in preventing precipitation: (a) A portion of the ionic calcium is removed by combination with protein and murexide. This removal may have a minimal effect, however, since the phosphate has a greater affinity for calcium than has protein or murexide. Apparently, phosphate has the greatest affinity for calcium, murexide an intermediate affinity, and protein the least affinity, since added phosphate takes calcium away from the murexide while protein has no influence on the murexide color in the presence of calcium. (b) Probably of more importance is the presence of dissimilar ions (e.g. sodium, chloride, potassium, hydroxyl), which inhibit the combination of calcium and phosphate ions but have no effect on the reverse reaction. Thus the binding of calcium by phosphate is effectively reduced. (c) Finally, the calcium concentration is lowered by addition of EDTA before sufficient time has elapsed for the frank precipitation to take place. There is no doubt that a considerable portion of the calcium in the solution is combined with phosphate, however, since considerably less calcium purpurate complex is formed with serum than with a standard having the same calcium concentration, the reduction in the complex formation being roughly proportional to the inorganic phosphate concentration.

From a consideration of these factors, it is apparent that the following conditions must prevail in order to obtain a valid direct titration of serum calcium with EDTA. All the calcium in the solution must be in a mobile equilibrium with the ionic calcium. With a mobile equilibrium, the steep portion of the titration curve faithfully points to the end point. The slope of the titration curve is diminished with increasing phosphate concentrations since larger portions of the calcium chelated by the EDTA are derived from the calcium bound with phosphate rather than from that bound to the indicator. This is borne out by the finding that the slope of the titration curve is, in a rough way, inversely proportional to the inorganic phosphate concentration. Secondly, the ionic strength of the solution must not change appreciably during the titration. With changes in the ionic strength, changes in the calcium and phosphate equilibriums will distort the end point. In the attempt to minimize the shift in ionic strength by incorporating a large amount of NaCl into the mixture and the titrant, another disturbing factor was uncovered: Even in the absence of phosphate, a high concentration of NaCl caused a shift of the apparent end point to higher values possibly due to an interference of calcium complex formation with EDTA.

The present method was designed with these factors in mind. Although fading of the indicator is minimal in the presence of 0.01 M NaOH, its use was abandoned since no way could be found for controlling the ionic strength of the solution. An NaOH concentration of 0.045 M is used so that the ionic strength of the final solution is derived, for the most part, from the base. Fading of the indicator is much slower than with the previous method (9) which utilizes 0.18 M NaOH. Also a quantity of NaCl was incorporated into the control, standard, and titrant of an ionic strength equivalent to about the median of the range that might be anticipated for serums from patients. When these reagents were used, even with the most severe electrolyte derangements of the patient, the ionic strength of the sample being titrated would vary only a few tenths of a per cent per milliliter of titrant added. The modified procedure has been used in the clinical laboratory for an extended period and no difficulties have been encountered. Phosphate, even at a high concentration, does not interfere with the titration. A serum sample was titrated with and without added phosphate. The serum itself contained 3.9 mg./100 ml. inorganic phosphorus, while the sample with added inorganic phosphate contained a total of 16.3 mg./100 ml. inorganic phosphorus, a phosphate level that might be encountered in severe uremia. The results are shown in Fig. 1. Although a considerable portion of the calcium conjugated with the phosphate, a mobile equilibrium was maintained since the two titrations were in agreement. This finding is in agreement with Kenny and Toverud (14).

The results shown in Table I indicate excellent agreement with those obtained by the Clark and Collip (15) modification of the Kramer and Tisdall (16) method. Although some investigators

TABLE I

SERUM CALCIUM VALUES

Se-rum no.	Method	Serum volume (ml.)	Calcium added (mg. %)	Number of detns.	Mean (mg. %)	Standard deviation (mg. %)	C.V.[a] (%)	Re-covery (%)
1	S[b]	2	0	5	10.10	0.265	2.67	
1	II	1	0	5	10.04	0.105	1.05	
1	I	0.5	0	1	10.1			
2	S[b]	2	0	3	10.20	0.100	1.05	
2	II	1	0	12	10.15	0.128	1.26	
3	II	1	0	4	10.69	0.067	0.63	
3	I	0.5	0	9	10.56	0.096	0.90	
4	II	1	0	2	9.6			
4	II	1	5.0	2	14.5			99.4
4	II	0.8[c]	0	1	7.8			101.7

[a] C.V. = coefficient of variation.

[b] Method S is the standard Clark-Collip procedure (15).

[c] This 0.8 ml. of serum was diluted to 1.0 ml. with saline. The concentration of calcium in the diluted serum was calculated.

object to the use of this method as a standard of reference, it is felt strongly by the author that any direct measurement on a complex mixture such as serum must be verified by an adequate isolation procedure. Incidentally, the excellent agreement obtained should dispel some of the doubts expressed concerning this old, but apparently reliable, procedure.

About equal reproducibility is obtained by methods I and II when the same sized cuvettes are used (19-mm.). The sensitivity is diminished, however, with cuvettes having a shorter light path. The reproducibility is of a lower order than was reported for the original procedure. It has been found that the end point derived graphically is not wholly independent of the indicator concentration, as was originally assumed. Therefore, a standard must be run with each set of determinations and the final result depends upon two measurements rather than one.[b] Each of the determinations reported in the table were carried out separately with a standard so that the true standard deviation could be calculated. The small shifts in the end points that occur with varying indicator concentrations are proportional, regardless of the calcium concentration in the sample being titrated, so that valid results are obtained for each titration when a standard is concurrently titrated with the same alkaline indicator solution. This is brought out in Table I by the satisfactory recovery obtained after the addition of calcium and after dilution.

[b] When duplicate determinations are made, it is advisable to include the replicate in a second titration rather than to titrate the duplicates with a single standard. The mean thereby will not be biased by the result obtained for the titration of a single standard. The author is indebted to Miss Mary L. Baumann for this suggestion.

The decision whether or not to run duplicate determinations in the clinical laboratory must be made by the clinical chemist and depends upon the reproducibility of the determination, on the work load, and on the personnel problem existing in the individual laboratory. Undeniably, the confidence in the results obtained by duplicate determinations for all chemical procedures is manyfold the confidence that may be placed on a single determination. Therefore, if practical, all clinical chemical determinations should be run in duplicate. However, it should be mentioned that the results obtained with a single determination by the direct method presented may be relied upon with considerably more confidence than the result obtained from a single determination by the isolation procedures. This is because the direct method eliminates the precipitation and washing steps, thus removing possible sources of gross errors (loss of precipitate, incomplete washing, etc.).

REFERENCES

1. Schwarzenbach, G., Biedermann, W., and Bangerter, F., Complexone VI. Neue einfache Titriermethode zur Bestimmung der Wasserhärte. *Helv. Chim. Acta.* **29**, 811–819 (1946).

2. Schwarzenbach, G., and Gysling, H., Metallindikatoren I. Murexid als Indicator auf Calcium- und andere Metall-Ionen. Komplexbildung und Lichtabsorption. *Helv. Chim. Acta.* **32**, 1314–1325 (1949).

3. Sobel, A. E., and Hanok, A., A rapid method for the determination of ultramicro quantities of calcium and magnesium. *Proc. Soc. Exptl. Biol. Med.* **77**, 737–740 (1951).

4. Buckley, E. S., Jr., Gibson, J. B., 2nd, and Bortolotti, T. R., Simplified titrimetric techniques for assay of calcium and magnesium in plasma. *J. Lab. Clin. Med.* **38**, 751–761 (1951).

5. Elliot, W. E., Volumetric determination of calcium in blood serum. *J. Biol. Chem.* **197**, 641–644 (1952).

6. Holtz, A. H., Directe titratie van calcium en magnesium in bloedserum. *Chem. Weekblad* **47**, 907–909 (1951).

7. Holtz, A. H., and Seekles, L., Direct titration of calcium in blood serum. *Nature* **169**, 870–871 (1952).

8. Flaschka, H. and Holasek, A., Eine neue Methode zur Bestimmung des Calciums im Blutserum. *Z. physiol. Chem.* **288**, 244–249 (1951).

9. Fales, F. W., A micromethod for the determination of serum calcium. *J. Biol. Chem.* **204**, 577–585 (1953).

10. Kibrick, A. C., Ross, M., and Rogers, H. E., Microdetermination of calcium in blood serum by direct titration. *Proc. Soc. Exptl. Biol. Med.* **81**, 353–355 (1952).

11. Dreven, B., Bernard, A., and Mandon, D., Microdosage du Ca^{++} sanguin utilisant l'acide éthylènediaminetétracétique et la murexide. *Compt. rend. soc. biol.* **147**, 1420–1424 (1953).

12. Williams, M. D., and Moser, J. H., Colorimetric determination of calcium with ammonium purpurate. *Anal. Chem.* **25**, 1414–1417 (1953).

13. Logan, M. A., and Taylor, H. L., Solubility of bone salt. *J. Biol. Chem.* **119**, 293–307 (1937).

14. Kenny, A. D., and Toverud, S. U., Noninterference of phosphate in an ethylenediaminetetraacetate method for serum calcium. *Anal. Chem.* **26**, 1059 (1954).

15. Clark, E. P., and Collip, J. P., A study of the Tisdall method for the determination of blood serum calcium with a suggested modification. *J. Biol. Chem.* **63**, 461–464 (1925).

16. Kramer, B., and Tisdall, F. F., A simple technique for the determination of calcium and magnesium in small amounts of serum. *J. Biol. Chem.* **47**, 475–481 (1921).

CEPHALIN-CHOLESTEROL FLOCCULATION TEST*

Submitted by: MARJORIE KNOWLTON, Walter Reed Army Institute of Research,
Walter Reed Army Medical Center, Washington, D. C.†

Checked by: BENNIE ZAK, Department of Pathology, Wayne State University
College of Medicine, Detroit, Michigan
GEORGE T. LEWIS and HELEN L. LANGEN, Department of Biochemistry, University of Miami School of Medicine, Coral Gables, Florida

Introduction

In 1938, Hanger (1) described a qualitative flocculation test, in which he mixed saline diluted serum with an emulsion prepared from sheep's brain lipid, cephalin, and cholesterol. After 24 and 48 hours, the absence or presence of flocculation was noted. On the basis of a series of 900 serums taken from normal individuals and at random from patients in medical wards, Hanger (2) concluded that flocculation did not occur with normal serum, but did with serum from patients with diseases of the liver parenchyma, that the degree of flocculation was an index of the liver impairment, and that flocculation would usually distinguish hepatogenous from obstructive jaundice.

In 1944, Neefe and Reinhold (3) demonstrated the effect of light and heat on the reaction. In order to avoid falsely positive reactions and for uniform results, it was necessary to protect the saline diluted serum and the mixture of diluted serum and antigen from overlong exposure to natural or artificial light. Furthermore, more reliable results were obtained at 20°–25°C. than at 37.5°C., at which temperature the rate and degree of flocculation increased.

Other modifications, aimed principally at obtaining a quantitative estimation of the reaction, have been made: serial dilutions of the serum with saline (4–7) to determine the point at which floccu-

* Based on the method of Hanger (1).
† Present address: Veterans Administration Hospital, Coatesville, Pa.

12

lation occurs; photometric measurement of the residual turbidity of the supernatant fluid (8) and of the flocculated material for its cholesterol content (9, 10). In an effort to obtain a more stable antigen emulsion, it has been suggested that an antigen prepared from cholesterol deoxycholate (11) be used and that the stock ether antigen solution be added to water at room temperature rather than at 65°–70°C. (6).

Principle

Certain saline diluted serums when mixed with cephalin-cholesterol antigen emulsion will cause flocculation and precipitation of the colloidal particles to form a globulin-cholesterol complex. In some instances, complete flocculation and precipitation occur, leaving a clear supernatant fluid.

Mechanism of the Test

In the original description of the test, Hanger (1) stated that the washed precipitate from a positive reaction contained a nitrogen-bearing constituent attached to the lipid surface. Electrophoretic and chemical separations of the protein fractions of normal and pathological serums have greatly advanced the understanding of the flocculation and turbidity reactions (12–17). A positive reaction of the cephalin-cholesterol test is attributed by Hanger to an increase of the gamma globulin fraction in such quantity as to prevent the flocculation-inhibiting action of albumin (which is generally abundantly present in nonhepatic disorders), to a decrease in albumin (as found in cirrhosis), or to an alteration of the inhibiting quality of albumin (as found in viral hepatitis). The flocculation of proteins in this system is a poorly understood physicochemical phenomenon dependent on several factors such as temperature, pH, dilution, and ion concentration (17).

Reagents

1. *Merthiolate, stainless, aqueous, 1 g./l. of water (1:1000).* Store in a dark reagent bottle. This reagent is used as a preservative and does not affect the reaction.

NOTE: Merthiolate may be omitted, according to one checker.

2. *Sodium chloride, 0.85%.* Dissolve 8.5 g. of reagent grade NaCl and 20 ml. of the merthiolate solution in a liter of water. For uni-

formity of results, prepare the solution in a 1-l. volumetric flask. If a precipitate forms in the reagent, discard.

3. Ether, anesthesia quality, Squibb. This grade is preferred, rather than anhydrous ether, because its moisture aids in solution of the antigen.

4. Cephalin-cholesterol antigen. Prepare from 100 mg. of sheep's brain cephalin and 300 mg. of cholesterol. The dried product is commercially available in vials.

WORKING SOLUTIONS

1. Stock solution. Add a quantity of ether (5 or 8 ml. as designated by the manufacturer) to the dried antigen in the vial. Stopper tightly with the screw cap provided with the vial and mix thoroughly by inversion. To obtain complete solution, it is advantageous to let the solution stand for a day before use.

NOTE: It has been stated that the stock solution will remain stable for "many months" if kept tightly stoppered. This writer has obtained satisfactory results when using antigen stored for three months at 20°–25°C. but has had no experience with solutions stored for a longer period. An assumption may be made that if an antigen is satisfactory when freshly prepared, it will remain stable for an indefinite period. (see Table I.) (according to Hanger).

2. Emulsion. Heat 35 ml. of freshly redistilled water to 65°–70°C. and add 1 ml. of the stock ether solution slowly with stirring. Heat the mixture at a temperature low enough to avoid foaming until the volume is approximately 30 ml. Cool to room temperature before use. The emulsion may be stored at refrigerator temperature (5°–10°C.) from 2 to 4 weeks.

NOTE: It was observed that when curdled material was present in the emulsion prepared according to Hanger, false positive reactions occurred. When the precipitate was removed from the emulsion by centrifugation, false positive reactions diminished without altering the quality of this reagent. It is, therefore, recommended that when a precipitate occurs it be removed by centrifugation. Data are provided below to establish the value of this additional step in the preparation of this reagent.

Glassware

1. Buret: 50-ml., with reservoir for easy refilling.

2. Pipets: Serological, 5-ml. and 0.2-ml. blowout.

3. Reaction tubes: 15-ml. centrifuge tubes or round-bottom tubes, 15 x 125 mm.

TABLE I

Effect of Age on Stock Ether Solution

Antigen no.	Age of stock (days)	Age of emulsion (days)	Serum controls						
			Neg.	Neg.	Neg.	Neg.	Pos.	Pos.	Pos.
1. (407)	1	2	0	0	0		3+	2+	
	30	1	0	±	±		2+	2+	2+
	64	Fresh	1+	2+	1+	2+			
2. (407)	1	2	0	0	±		2+	2+	
	30	1	0	0			2+		
	64	2	0	0					
3. (414)	22	Fresh	0	1+	0		2+	2+	
	88	Fresh	0	0			3+		
4. (414)	5	Fresh	0	0	0	0	2+		
	46	Fresh	0	0			3+		
5. (414)	2	Fresh	0	0	0		3+		
	27	Fresh	0	0			3+		
6. (414)	2	Fresh	0	0			2+		
	27	Fresh	0	1+			4+ (gel. floc.)[a]		
7. (414)	4	Fresh	0	0	0				
	25	Fresh	0	1+			4+ (gel. floc.)[a]		
8. (414)	4	Fresh	0	0			1+	3+	
	30	Fresh	2+	2+	1+		2+	3+	

[a] Gelatinous flocculation: a characteristic flocculation frequently observed in falsely positive floccula-tions.

NOTE: The advantage of the conical centrifuge tube is that one may estimate the amount of precipitate, if any is present. However, because of the large breakage factor of the conical tube, the round-bottom tube is recommended. With practice, the reading of the precipitate may be made easily.

All glassware must be carefully cleaned: ion concentration and pH are critical in flocculation reactions; therefore, foreign particles and acid and alkali cleaning agents must be completely removed by thorough rinsing with tap and distilled water.

Procedure

1. Withdraw a blood sample from the patient in the fasting state. Let the blood clot and centrifuge to obtain clear serum.

NOTE: For uniform results, use fresh serum whenever possible. If necessary, serum may be refrigerated overnight. Frozen serum may be used if it is quickly thawed at 45°C. to avoid alteration of the proteins.

2. With each series of tests include a negative and a positive control serum and an antigen control (4 ml. of saline and 1 ml. of emulsion).

3. Prepare a 1:21 dilution of serum with 0.85% NaCl solution by adding 4 ml. of saline to 0.2 ml. of serum in the reaction tube. Mix thoroughly by tapping (10–12 times).

4. Add 1 ml. of the emulsion to each tube and mix thoroughly by tapping (15–20 times).

5. Cover the tubes and place in a dark cabinet.

6. After 24 hours read the reaction:

    a. No flocculation or precipitation: negative.

    b. Minimal precipitation and/or flocculation: 1+.

    c. Definite precipitation and flocculation: 2+.

    d. Definite precipitation and residual turbidity in the supernatant fluid: 3+.

    e. Complete precipitation and clear supernatant fluid: 4+.

NOTE: Standardize antigens with a new lot number by running several negative and positive control serums with the new antigen and with the antigen being used.
If the proportions are kept the same, smaller samples of serum may be used.

Comments

The cephalin-cholesterol flocculation test, while simple to perform, unfortunately requires the use of an antigen emulsion which is unstable. There is variation in sensitivity of single vials of antigen and of different lots of antigen as well as between antigens prepared by the leading manufacturers. Occasionally false negative and positive reactions occur because of poor control of the antigen.

Flocculation is a phenomenon dependent on several factors, e.g. temperature, pH, dilution, and surface charges. Variations of any of these may change the action (17). Certain precautionary measures have been suggested in order to maintain uniform results:

1. Use of carefully cleaned glassware.

2. Use of freshly redistilled water to remove bacteria and other impurities, such as electrolytes.

3. Use of fresh serum to avoid alterations of the proteins.

4. Protection of the diluted serum and the reaction mixture from light and heat (3). Improvement of the method of preparing the dried antigen may have overcome the photosensitivity formerly observed (18). Sufficient ripening of cephalin as suggested by Mateer et al. (19) may have eliminated this effect as well as reduced the number of false positive reactions. In a small series of tests comparing reactions exposed to light and heat with reactions in the dark, the tests exposed to fluorescent light differed only slightly from the "dark" tests, while those exposed to sunlight and to a tungsten lamp showed changes from negative to positive or increased positivity (see Table II).

5. Use of merthiolate (1:1000 or 1:10,000 dilutions) as a preservative in the antigen emulsion is not an effective procedure. False positive reactions occur with antigens treated with merthiolate. Emulsions prepared at the same time from the same stock ether solution, one with added merthiolate and the other untreated, will deteriorate in the same length of time (see Table III).

TABLE II

THE EFFECT OF LIGHT AND HEAT ON THE FLOCCULATION REACTION

Serum	Dark	Fluorescent light	Tungsten lamp	Sunlight
1	0	0		2+
2	0	1+		1+
3	0	0		2+
4	0	0		1+
5	1+	1+		2+
6	1+	1+		1+
7	0		1+	
8	2+		2+	
9	0		1+	
10	0		1+	
11	2+	3+		3+
12	2+	2+		3+
13	0	0		2+
14	2+	3+		3+
15	3+	3+		4+
16	3+	3+		4+

TABLE III

STABILITY OF EMULSION AND EFFECT OF CENTRIFUGATION

Antigen (Emulsion) no.	Age of emulsion (days)	Controls						
		Anti-gen	Neg.	Neg.	Neg.	Neg.	Pos.	Pos.
1[b]	1	1+	1+				2+	1+
	1[c]	0	0				2+	1+
	2	1+	2+				3+	3+
	2[c]	0	±				1+	1+
2[d]	1	1+	1+				2+	1+
	1[c]	0	0				2+	1+
	2	2+	1+				3+	2+
	2[c]	±	±				3+	2+
	4	1+	2+	2+			2+	
	4[c]	0	±	0			2+	
3[b]	6	1+	2+[e]				1+	3+
	6[c]	0	2+[e]				1+	3+
	7	1+	1+	1+				
	7[c]	0	0	0				
	16	1+	1+[f]	1+	1+		3+	
	16[c]	0	0[f]	0	0		3+	
4	Fresh	1+	1+	0	1+		3+	
	Fresh[c]	0	0	0	0		3+	
	4	1+	1+	2+	1+		3+	
	4[c]	0	0	0	0		3+	
	24	0	1+	2+			Gel. floc.	
	24[c]	0	3+	1+			Gel. floc.	
	26[c]	0	2+	1+			Gel. floc.	
5[a]	Fresh	1+	0		1+	0	3+	
	Fresh[c]	0	0		0	0	3+	
	24	1+	3+		3+		3+	
	26[c]	0	0		2+		2+	
6	Fresh	1+	0		1+	0	3+	
	Fresh[c]	0	0		0	0	3+	
	24	0	Gel. floc.		1+		Gel. floc.	
	24[c]	0	3+		1+		Gel. floc.	
	26[c]	0	2+		1+		Gel. floc.	

Antigen (Emulsion) no.	Age of emulsion (days)	Controls						
		Antigen	Neg.	Neg.	Neg.	Neg.	Pos.	Pos.
7 [b]	2	0	1+	1+	1+	1+	3+	
	2 [c]	0	0	0	0	0	3+	
	30 [c]	0	±	1+	1+		4+	
8	2	0	0	0	1+	1+	3+	
	2 [c]	0	0	±	±	±	3+	
	30 [c]	0	±	1+	1+		4+	
9	Fresh	0	0	0				
	6	±	±	2+			4+	
	29 [c]	0	0	±	±		4+	
	53 [c]	0	Gel. floc.	Gel. floc.			Gel. floc.	
10 [b]	Fresh	0	0	0				
	8 [c]	0	0	0	±		3+	
	24 [c]	0	0	±	±		4+	
	58 [c]	0	4+	4+	4+		4+	
11	Fresh	1+	1+	1+			3+	3+
	Fresh [c]	0	0	0			3+	3+
	18 [c]	0	1+				2+	3+
12	Fresh	1+	1+	±			3+	3+
	Fresh [c]	0	0	0			3+	3+
	5 [c]	0	0	0			3+	4+
	18 [c]	0	3+				3+	3+

[a] Preservative added (1:10,000 merthiolate).
[b] Preservative added (1:1000 merthiolate).
[c] Emulsion centrifuged.
[d] Stock ether solution added to distilled water at 25°C.
[e] Serum frozen and thawed twice.
[f] Serum refrigerated 3 days.

Modification for the Preparation of the Antigen Emulsion

In an attempt to determine the cause of a series of false positive reactions, the emulsion was centrifuged to remove the precipitated "curdled" material present in the emulsion. The same serum tested with milky smooth supernatant emulsion showed no false positive reactions.

Many trials have been made by the submitter, using stock solutions from two different lots of antigen which had been discarded as unsatisfactory. The emulsions were prepared according to the Hanger directions. After cooling to room temperature, the emulsions were mixed with a rotary motion and examined over a strong light. If aggregations of precipitate occurred, the emulsions were centrifuged at high speed for 20 to 25 minutes. A precipitate was thrown to the bottom of the tube and a button of lipid material rose to the top. The button was pushed aside gently and the milky, smooth supernatant emulsion removed by aspiration. Comparative tests were made with the freshly prepared antigen emulsion before and after centrifugation. False positive reactions were obtained with the negative control serums before centrifugation. The centrifuged emulsion, however, gave no false positive reactions with the same serums. Centrifugation did not interfere with the flocculation of abnormal serum, nor did it prevent the deterioration with age of the emulsion (see Table III).

Variation in Healthy Persons

The sensitivity of the reagent determines to some extent the number of positive responses in a healthy person. In general, however, 2% of healthy persons, 10% of hospital patients, 50 to 80% of patients with cirrhosis and 85% of patients with hepatitis give a response of 2 + or more with the cephalin-cholesterol flocculation test.

Conclusion

The cephalin-cholesterol flocculation test is simple to perform and requires a minimum of apparatus. It is an empirical test of disease of the liver and, as originally outlined by Hanger, continues to be a widely used and useful test.

A suggestion for the recovery of unsatisfactory antigen emulsion by centrifugation is offered.

REFERENCES

1. Hanger, F. M., The flocculation of cephalin-cholesterol emulsions by pathological sera. *Trans. Assoc. Am. Physicians* **53,** 148–151 (1938).
2. Hanger, F. M., Serological differentiation of obstructive from hepatogenous jaundice by flocculation of cephalin-cholesterol emulsion. *J. Clin. Invest.* **18,** 261–269 (1939).

3. Neefe, J. R., and Reinhold, J. G., Photosensitivity as a cause of falsely positive cephalin-cholesterol flocculation tests. *Science* **100,** 83–85 (1944).
4. Bruger, M., Fractional cephalin-cholesterol flocculation in hepatic disease. *Science* **97,** 585–586 (1943).
5. Mirsky, I. M., and von Brecht, R., The fractional cephalin-cholesterol flocculation test. *Science* **98,** 499–500 (1943).
6. Frisch, A. W., and Quilligan, J. J., Jr., Modified cephalin-cholesterol test in the study of hepatic disease. *Am. J. Med. Sci.* **212,** 143–152 (1946).
7. Makari, J. G., A non-specific resistance factor in the albumin residue as revealed by the serial cephalin flocculation test. *Nature* **160,** 201–203 (1947).
8. Kibrick, A. C., Rogers, H. E., and Skupp, S. J., A photometric method for the precise estimation of cephalin-cholesterol flocculation. *Am. J. Clin. Pathol.* **22,** 698–702 (1952).
9. Saifer, A., A method for the quantitative determination of the cephalin-cholesterol flocculation. *J. Clin. Invest.* **27,** 737–744 (1948).
10. Jennings, E. R., Cherney, P., and Zak, B., Spectrophotometric method for determination of cephalin-cholesterol flocculation test. *Am. J. Clin. Pathol.* **23,** 1173–1178 (1953).
11. Steinberg, A., Cholesterol-desoxycholic acid: a stable antigen for use in a flocculation test for liver dysfunction. *J. Lab. Clin. Med.* **34,** 1049–1056 (1949).
12. Maclagan, N. F., and Bunn, D., Flocculation tests with electrophoretically separated serum proteins. *Biochem. J.* **41,** 580–586 (1947).
13. Kabat, E. A., Hanger, F. M., Moore, D. H., and Landow, H., Relation of cephalin flocculation and colloidal gold reactions with serum proteins. *J. Clin. Invest.* **22,** 563–568 (1943).
14. Moore, D. B., Pierson, P. S., Hanger, F. M., and Moore, D. H., Mechanism of the positive cephalin-cholesterol flocculation reaction in hepatitis. *J. Clin. Invest.* **24,** 292–295 (1945).
15. Hanger, F. M., Abnormalities in the globulin component of serum as demonstrable by the cephalin flocculation test. *Trans. Assoc. Am. Physicians* **60,** 82–85 (1947).
16. Cohn, E. J., A system for the separation of the components of human blood. *J. Am. Chem. Soc.* **72,** 465–474 (1950).
17. Saifer, A., Protein flocculation reactions. *Am. J. Med.* **13,** 730–743 (1952).
18. Reinhold, J. G., Chemical evaluations of liver function. *Clin. Chem.* **1,** 351–421 (1955).
19. Mateer, J. G., Baltz, J. I., Marion, D. F., and MacMillan, J. M., Liver function tests. *J. Am. Med. Assoc.* **121,** 723–728 (1943).

CHLORIDE*

Submitted by: Bernard Klein, Biochemistry Laboratory, Veterans Administration Hospital, The Bronx, New York

Checked by: Margaret Vanderau, Chemical Laboratory, Presbyterian Hospital, Philadelphia, Pennsylvania

Cecelia Riegel, Chemical Laboratory, Lankenau Hospital, Philadelphia, Pennsylvania

Introduction

The following procedure for the determination of chloride was developed to overcome the shortcomings of the original argentimetric-adsorption-indicator method of Saifer and Kornblum (1) and its subsequent modifications (2, 3). In its present form (4), an aliquot of a barium hydroxide-zinc sulfate deproteinized serum centrifugate (5) is titrated with standardized silver nitrate; dichlorofluorescein is used as an adsorption indicator.

Principle

A suspension of silver chloride in the presence of excess chloride ion acquires a negative surface charge owing to the adsorption of the latter $(AgCl:Cl^-)^-$. When the equivalence point is reached, as in an argentimetric titration, some of the excess silver ion added will be adsorbed by the AgCl particles $(AgCl:Ag^+)^+$. Loosely held negative ions like NO_3^- in the presence of more strongly adsorbed dichlorofluorescein ions will be readily replaced on the precipitate surface by the indicator ions. The color change does not take place in the solution, but on the surface of the precipitate.

In practice, when a drop of dilute dichlorofluorescein solution is added to the chloride solution, the color of the indicator remains the same as in water. The addition of silver nitrate precipitates AgCl, colloidally suspended in the liquid. Just before the equivalence point, the precipitate flocculates. With the introduction of the first excess Ag^+, the dichlorofluorescein ions are adsorbed on the AgCl surface and the suspended matter turns pink.

* Based on the method of Franco and Klein (4).

22

Reagents

1. Silver nitrate solution, 0.1 N. Dissolve 16.989 g. of dried, analytical grade crystals in water and dilute to 1 l. When stored in a brown bottle it is stable.

2. Silver nitrate solution, 0.02 N. Prepare by diluting 200 ml. 0.1 N solution to 1 l. When stored in a brown bottle it is stable.

3. Sodium chloride standard, 0.0200 N. Dissolve 0.5845 g. of dried, analytical grade crystals in water and dilute to 500 ml.

4. Standardization of silver nitrate. To 1.00 ml. of the NaCl standard (0.02 meq.) add 1 drop of indicator solution and titrate with the 0.02 N silver nitrate, using a microburet, to the first pink.

5. Dichlorofluorescein indicator, 0.05%. Dissolve 0.05 g. dichlorofluorescein (Distillation Products Industries No. 373) in 100 ml. 70% ethyl alcohol.

6. Zinc sulfate, 5%. Dissolve 50 g. reagent grade $ZnSO_4.7\ H_2O$ in water and dilute to 1 l.

7. Barium hydroxide, 0.3 N. Dissolve 47.3 g. reagent grade $Ba(OH)_2.8\ H_2O$ in about 900 ml. water and dilute to 1 l. Allow to stand overnight to permit the precipitated carbonate to settle and then titrate the supernatant against the zinc sulfate solution. Five milliliters zinc sulfate solution is diluted to 30 ml. with water and *slowly* titrated with the barium hydroxide solution (with phenolphthalein as the indicator) until 1 drop turns the solution a faint pink. About 4.7–4.8 ml. should be used. Otherwise the solutions should be adjusted accordingly. The barium hydroxide solution should be siphoned into and kept in a bottle protected from atmospheric carbon dioxide by a soda lime tube.

For use: The barium hydroxide solution is diluted in the ratio 2 volumes to 5 volumes water.

NOTE 1: In the submitter's laboratory the diluted solution is stored in a 3-l. bottle fitted with a 50-ml. automatic buret equipped with a soda lime tube.

Procedure

Determination of Serum Chloride

To 1.00 ml. fresh serum in a 16 x 150-mm. test tube add 7.0 ml. of diluted $Ba(OH)_2$ solution, then add 2.0 ml. $ZnSO_4$ solution while shaking the tube. Stopper the tube, shake well, and centrifuge for 10 minutes at 2000 r.p.m.

Note 2: The centrifugate may also be used for the determination of "true" glucose (6). If urease treatment precedes precipitation of proteins, urea N and glucose may be measured in the supernatant (6).

Place 2.00 ml. of the centrifugate in a 16 x 150-mm. test tube, add 1 drop of indicator solution and titrate with the standardized $AgNO_3$ until the first pink is seen throughout the solution.

Note 3: A 5.0-ml. buret graduated in 0.01 ml. is recommended.
Note 4: The titration should not be carried out in direct sunlight since silver chloride with adsorbed indicator is extremely light-sensitive. The red color changes to gray and then to black in strong light.

Determination of Spinal Fluid Chloride

Treat spinal fluid as serum and titrate the centrifugate as described for the determination of serum chloride. (See note 2.)

Note 5: This procedure does not differentiate chloride from bromide because total halide is titrated.

Calculation

Milliequivalents Cl^-/liter = Milliliters $AgNO_3$ used in titration $\times$ F

$$F = \frac{100}{\text{Milliliters } AgNO_3 \text{ used in standardization}}$$

Normal Values

Chloride in serum from healthy persons varies from 98 to 105 meq. per liter.
Chloride in spinal fluid varies from 125 to 135 meq. per liter.

TABLE I

DETERMINATION OF CHLORIDE IN POOLED SERUM

Sample	Schales and Schales (7) (meq./l.)	Adsorption-indicator method (meq./l.)	Difference (meq./l.)
1	101.5	100.6	−0.9
2	100.5	100.6	+0.1
3	101.5	101.6	+0.1
4	101.5	101.6	+0.1
5	101.5	100.6	−0.9

Comments

1. The procedure gives values in excellent agreement with the mercurimetric-diphenylcarbazone procedure (7) (see Table I).
2. Replicate titrations were reproducible within 0.6%.
3. Recovery studies on pooled normal serums indicated a mean of 101.2% recovery with a range of 100.5 to 101.8%.
4. The procedure is best carried out under those conditions wherein the precipitate separates as a flocculated colloid. Large amounts of neutral salts obscure the end point because they exert their flocculating effect before the stoichiometric end point is reached. This is probably the case in the titration of urinary chloride, although successful titrations of Hagedorn-Jensen filtrates have been reported (2).

REFERENCES

1. Saifer, A., and Kornblum, M., Determination of chlorides in biological fluids by use of adsorption indicators. *J. Biol. Chem.* **112**, 117–122 (1935).
2. Saifer, A., and Hughes, J., Determination of chlorides in biological fluids by use of adsorption indicators. *J. Biol. Chem.* **129**, 273–281 (1939).
3. Saifer, A., Hughes, J., and Weiss, E., Determination of chlorides in biological fluids by use of adsorption indicators. *J. Biol. Chem.* **146**, 527–535 (1942).
4. Franco, V., and Klein, B., The microdetermination of chlorides in serum and spinal fluid. *J. Lab. Clin. Med.* **37**, 950–954 (1951).
5. Somogyi, M., Determination of blood sugar. *J. Biol. Chem.* **160**, 69–73 (1945).
6. Reinhold, J. G., Glucose. *In* "Standard Methods of Clinical Chemistry" (M. Reiner, ed.), Vol. 1, p. 67. Academic Press, New York, 1953.
7. Schales, O., Chloride. *In* "Standard Methods of Clinical Chemistry" (M. Reiner, ed.), Vol. 1, p. 37. Academic Press, New York, 1953.

PERTINENT REVIEWS AND BOOKS

1. Gamble, J. L., "Chemical Anatomy, Physiology and Pathology of Extracellular Fluid," 5th ed. Harvard Univ. Press, Cambridge, Massachusetts, 1947.
2. Weisberg, H. F., "Water, Electrolyte and Acid-Base Balance." Williams & Wilkins, Baltimore, Maryland, 1953.
3. Elkinton, J. R., and Danowski, T. S., "The Body Fluids," Williams & Wilkins, Baltimore, Maryland, 1955.

CHOLESTEROL IN SERUM*

Submitted by: LIESE L. ABELL and BETTY B. LEVY, Columbia University, College of Physicians and Surgeons, Goldwater Memorial Hospital, New York, New York.

BERNARD B. BRODIE, Department of Chemical Pharmacology, National Institutes of Health, Bethesda, Maryland.

FORREST E. KENDALL, Department of Biochemistry, Columbia University, New York, New York.

Checked by: A. KAPLAN and S. JACQUES, Department of Biochemistry, Medical Research Institute, Michael Reese Hospital, Chicago, Illinois

JOHN REINHOLD, Pepper Laboratory of Clinical Medicine, Hospital of the University of Pensylvania, Philadelphia, Pennsylvania

Introduction

The determination of total cholesterol in serum has assumed an added significance in recent years because of the possible implication of cholesterol in the etiology of arteriosclerosis. There is need for a method that is sufficiently flexible to yield equally good results in clinical laboratories where only occasional analyses are made and in research laboratories where highly trained personnel is available to carry out a large number of determinations each day.

Principle

The simplified but precise method described in this paper involves (a) treatment of the serum with alcoholic potassium hydroxide to liberate the cholesterol from the lipoprotein complexes and to saponify the cholesterol esters; (b) extraction of the cholesterol into a measured volume of petroleum ether after dilution of the alcoholic solution with water; and (c) measurement of the cholesterol in an aliquot of the petroleum ether layer by means of the Liebermann-Burchard color reaction. Details of the method described in this paper have been borrowed freely from existing methods. The

---

* Based on the method of Abell, Levy, Brodie, and Kendall (2).

26

work of Sperry and Brand (1) has been especially valuable in furnishing a basis for our method (2).

Since substances other than cholesterol may give color with the Liebermann-Burchard reagent, the specificity of the method was assayed by the countercurrent distribution technique (3). More than 99% of the material in serum determined by this method is shown to be cholesterol.

Reagents

1. Absolute ethyl alcohol, redistilled.

2. Petroleum ether, b.p. 68°C., redistilled. The use of a high-boiling fraction minimizes errors by evaporation. However, this is not critical, and with care any redistilled petroleum ether may be used.

3. Acetic acid, reagent grade.

4. Sulfuric acid, reagent grade.

5. Acetic anhydride, reagent grade, free from HCl.

6. Potassium hydroxide solution, 33% (w/w). Dissolve 10 g. of reagent grade KOH in 20 ml. of water.

7. Alcoholic potassium hydroxide solution. Make immediately before using by adding 6 ml. of 33% KOH to 94 ml. of the absolute alcohol.

8. Standard cholesterol solution (0.400 mg./ml.). Dissolve 100.0 mg. of cholesterol (recrystallized four times from absolute alcohol and dried to constant weight) in absolute alcohol and make to 250 ml. with this solvent.

9. Modified Liebermann-Burchard reagent. Chill 20 volumes of acetic anhydride to a temperature lower than 10°C. in a glass-stoppered container, add 1 volume of concentrated sulfuric acid, and keep the well-shaken mixture cold for 9 minutes. Add 10 volumes of glacial acetic acid and warm to room temperature. The reagent should be used within 1 hour.

Procedure

Measure 0.50-ml. samples of serum or plasma into 25-ml. glass-stoppered centrifuge tubes, and add 5 ml. of alcoholic KOH to each tube. Stopper the tubes, shake well, and then warm in a water bath at 37°–40°C. for 55 minutes. After cooling to room temperature, add 10.0 ml. of petroleum ether and mix well with the contents of each tube. Add 5 ml. of water and shake the tubes vigorously for

1 minute. Then centrifuge at slow speed for 5 minutes or until the emulsion breaks and two clear layers appear. A suitable aliquot of the petroleum ether layer is transferred to a small, dry bottle.

NOTES: The aliquot taken should contain between 0.15 and 0.60 mg. of cholesterol. The use of 0.5-ml. samples of serums in the range of 150 to 300 mg. per 100 ml. permit duplicate 4-ml. aliquots.

It has been found convenient to use narrow-neck, 1-oz. medicine bottles at this point. Twelve to fourteen of these bottles can be fitted into a 4 x 5-inch wire basket for the remainder of the procedure.

A checker has suggested reducing the volumes used above to: serum—0.2 ml.; KOH—2 ml.; petroleum ether—5 ml.; water—2 ml.; centrifuge tubes—12-ml.; aliquots for color—2 ml. or 4 ml.

The analysis can be interrupted at this stage, and the bottles stoppered and kept until it is convenient to complete the procedure. Slow oxidation of the cholesterol occurs if the dried residues are permitted to stand.

Evaporate the petroleum ether by placing the bottles in a water bath at 60°C. and blowing a gentle stream of air into them. After cooling to room temperature, stopper the bottles with clean, dry corks. Store until a suitable time or prepare for color development with the Liebermann-Burchard reagent.

Prepare standards for inclusion in each series of determinations. This is most conveniently done by running the standard through the procedure along with the samples. Mix duplicate 5.00-ml. samples of the standard cholesterol solution (0.400 mg./ml.) and 0.3 ml. of 33% KOH solution in 25-ml. glass-stoppered centrifuge tubes and warm for 55 minutes at 37°–40°C. Add 10.0 ml. of petroleum ether and 5 ml. of water and vigorously shake the tubes for 1 minute. After centrifugation, measure 1.00-, 2.00-, and 3.00-ml. samples of the petroleum ether layer into 1-oz. bottles and evaporate to dryness to provide standards equivalent to 0.2, 0.4, and 0.6 mg. of cholesterol.

Arrange the bottles containing the dry residues from the samples and the standards in a wire basket so that a set of standards containing 0.2, 0.4, and 0.6 mg. of cholesterol appears at the beginning and another set at the end of the series. Place a clean, empty bottle at the beginning to receive the blank and set the samples in a water bath at 25°C. Start a stop-watch or timer and add 6.00 ml. of the modified Liebermann-Burchard reagent first to the empty bottle and then at regular intervals of time to the other samples. Carefully wash down the entire inner surface of the bottle with the reagent. Cork the bottles tightly, shake, and return to the bath. Read the optical density of each sample at regular intervals against the blank in a photoelectric colorimeter at 620 mμ, 30–35 minutes after the reagent has been added. Avoid intense light during color development. The usual laboratory lighting has little influence on the color.

Calculation of Results

The optical density equivalent to 1 mg. of cholesterol is calculated from the readings of the standards.

$$\frac{\text{Optical density of standard}}{\text{Milligrams cholesterol in standard}} = S$$

The S value for all the standards should agree within 4%.

NOTE: Occasionally the standards give optical densities that deviate from Beer's law. This behavior seems to be associated with some abnormality in the preparation of the Liebermann-Burchard reagent. When this occurs, use the standard giving the reading closest to each sample for calculating the cholesterol content of that sample.

Use the average of all the values for calculating the cholesterol content of the samples.

$$\frac{\text{Optical density of unknown}}{S} \times \frac{10}{\text{volume of petroleum ether aliquot}}$$

$$\times \frac{100}{\text{Volume of serum sample}} = \text{Milligrams cholesterol per 100 ml.}$$

Results and Discussion

The concurrent standards are an essential part of the method. They automatically correct for several factors which may give rise to considerable error. In calculating the results, the assumption is made that the volume of the petroleum ether layer is 10 ml. and that it contains all the cholesterol in the sample. Actually both these values are dependent upon the properties of the specific lots of solvents used and upon the temperatures at which the solvents are equilibrated. With different samples of solvents, the volume of the petroleum ether layer has been found to vary between 10.0 and

10.4 ml. and to contain between 95 and 100% of the total cholesterol in the sample. Since changes in these values affect the standards
and the samples alike, they cancel out and may be disregarded.

The internal standards serve as a check on each batch of Liebermann-Burchard reagent. Even though the reagent is always prepared in the same way, anomalous readings are occasionally obtained. In 100 consecutive routine analyses, the optical density of
the standards[1] averaged 0.823 per milligram cholesterol, with a
standard deviation of 0.023. The variation in readings was between
0.760 and 0.870. If this average figure is used to calculate the cholesterol content of a serum actually containing 200 mg.% of cholesterol, values ranging between 187 and 214 mg.% will be obtained
with different batches of reagent. This variation is three times that
found when the cholesterol is calculated from the internal standards
included in each run. The same variation in readings on the standard
is found in the Schoenheimer-Sperry method, where the color is
developed with known amounts of cholesterol which have not been
subjected to any intermediate procedure. In 100 consecutive routine
analyses, the optical density of the standards averaged 0.847 per
milligram cholesterol with a standard deviation of 0.023. The variation in readings was between 0.807 and 0.906.

The ratio between the optical densities of the standards in the
two methods is 0.97. This value represents the extent of recovery
of known amounts of cholesterol subjected to warming with alcoholic KOH and then extracted into petroleum ether.

To check the reproducibility of the method, 16 replicate 0.5-ml.
samples of the same human serum were divided into groups of four.
The first group was analyzed directly and the others after the addition of 0.20, 0.40, and 0.60 mg. of cholesterol, respectively (see
Table I). In each case, 4-ml. aliquots of the petroleum ether layer
were taken for color development. Four standards containing 2.0
mg. of cholesterol were run at the same time and 1-, 2-, and 3-ml.
samples of the petroleum ether layer were taken for color development. The optical density of each sample was measured at 620 mμ
exactly 30 minutes after the addition of the modified Liebermann-
Burchard reagent. The results are given in Table II. It will be seen
that the variation found in analyzing replicate samples is the same

[1] The standards were measured in matched 125 x 15-mm. Pyrex glass test tubes
in a spectrophotometer.

TABLE I

RECOVERY OF CHOLESTEROL ADDED TO 0.5 ML. OF SERUM[a]

Added cholesterol (mg.)	Optical density[b]	Total cholesterol (mg.)	Cholesterol recovered (% recovery)
0.00	0.359	1.06	
0.00	0.357	1.05	
0.00	0.354	1.04	
0.00	0.352	1.04	
0.20	0.417	1.23	98
0.20	0.426	1.25	100
0.20	0.429	1.26	101
0.20	0.429	1.26	101
0.40	0.491	1.44	96
0.40	0.498	1.46	101
0.40	0.498	1.46	101
0.40	0.495	1.45	100
0.60	0.561	1.65	100
0.60	0.565	1.66	101
0.60	0.585	1.72	104[c]
0.60	0.565	1.66	101

[a] Cholesterol content of serum sample, average 1.05 mg.; $\sigma = 0.01$.
[b] 4-ml. aliquots of the petroleum ether layer were taken for color development.
[c] Not included in the average.

TABLE II

OPTICAL DENSITY STANDARDS[a]

0.20 mg. Cholesterol		0.40 mg. Cholesterol		0.60 mg. Cholesterol	
Observed	Per milligram cholesterol	Observed	Per milligram cholesterol	Observed	Per milligram cholesterol
0.172	0.860	0.337	0.843	0.505	0.841
0.174	0.870	0.337	0.843	0.509	0.848
0.171	0.855	0.337	0.843	0.509	0.848
0.172	0.860	0.342	0.855	0.509	0.848

[a] Average optical density per milligram of cholesterol, 0.851; $\sigma = 0.009$.

as that found in determining the optical density of the standards. The accuracy of the method is limited by the reproducibility of duplicate spectrophotometer readings.

Eighteen samples of human serum were analyzed, both by this method and by the Schoenheimer-Sperry method (4). The agreement

TABLE III

COMPARISON OF SERUM CHOLESTEROL LEVELS

Serum no.	Proposed method duplicate determinations		Method of Schoen-heimer and Sperry
	mg./ml.	*mg./ml.*	*mg./ml.*
1	2.19	2.16	2.10
2	2.45	2.41	2.41
3	2.17	2.14	2.13
4	2.27	2.27	2.25
5	2.02	1.98	2.08
6	2.51	2.51	2.52
7	3.08	3.07	3.01
8	1.73	1.70	1.81
9	1.68	1.73	1.65
10	1.65	1.57	1.63
11	1.98	1.98	2.07
12	2.00	1.93	2.01
13	1.90	1.94	1.97
14	2.03	2.04	2.11
15	2.16	2.12	2.19
16	2.73	2.69	2.79
17	1.72	1.69	1.71
18	2.39	2.43	2.56

between the two methods was very good (see Table III).[2] In making these comparisons, all steps in both procedures were meticulously carried out. Several hundred samples have been routinely analyzed by both methods. In the routine laboratory, where many samples are being analyzed at one time, this method tends to give higher values than the Schoenheimer-Sperry method. The difference is not significant when the cholesterol level is below 300 mg./100 ml. but may occasionally become as large as 20% when high levels are encountered. In these cases, repeat determinations by both methods usually confirm the higher values. It is believed that these discrepancies are caused by failure to obtain complete hydrolysis of the esters in the Schoenheimer-Sperry determination. The procedure in a routine laboratory tends to become mechanical. Under these con-

[2] Two of the checkers checked this point and found it to be true. Five pools when analyzed 6–10 times showed the mean values, by both methods, to agree within 1.3%.

ditions occasional failure to achieve complete mixing of the strong KOH solution with the serum extract would result in low values with the Schoenheimer-Sperry method. Incomplete saponification does not affect the values obtained with the new method. With the modified Liebermann-Burchard reagent the color develops a little faster with an unsaponified sample, but the same maximum is reached in 30 minutes as that obtained with a completely saponified sample.[3] Saponification of the samples, while not essential for color development in the Liebermann-Burchard reaction, is needed to permit complete extraction of the sterol into the petroleum ether layer.

The specificity of the method has been tested by the counter-current distribution technique. Results of the application of this fractionation to two normal and to two hypercholesterolemic human serums indicate that more than 99% of the material determined by the analytical procedure is actually cholesterol.

REFERENCES

1. Sperry, W. M., and Brand, F. C., The Colorimetric determination of cholesterol. *J. Biol. Chem.* **150,** 315–324 (1943).
2. Abell, L. L., Levy, B. B., Brodie, B. B., and Kendall, F. E., Simplified method for the estimation of total cholesterol in serum and demonstration of its specificity. *J. Biol. Chem.* **195,** 357–366 (1952).
3. Craig, L. C., Golumbic, C., Mighton, H., and Titus, E., Identification of small amounts of organic compounds by distribution studies. III. Use of buffers in counter-current distribution. *J. Biol. Chem.* **161,** 321–332 (1945).
4. Schoenheimer, R., and Sperry, W. M., A micromethod for the determination of free and combined cholesterol. *J. Biol. Chem.* **106,** 745–760 (1934).

[3]Kendall, F. E., unpublished observation.

TOTAL FATTY ACIDS IN STOOL*

Submitted by: J. H. VAN DE KAMER, Central Institute for Nutrition Research,
T. N. O., Utrecht, The Netherlands.
Checked by: MARSCHELLE H. POWER, Biochemistry Laboratory, Mayo Clinic,
Rochester, Minnesota
DAVID A. TURNER, Surgical Research Metabolic Laboratory, George-
town University Hospital, Washington, D. C.

ntroduction

The determination of the fatty acid content of feces may be
carried out before or after drying of the feces. There are some objections to the latter method: (a) Drying takes a long time. (b)
During drying conversions may take place by which the ratio of
split fat to unsplit fat might be changed. (c) During drying, volatile
fatty acids may evaporate. These objections are eliminated by the
determination of fat in wet feces.

Several authors have published methods for the determination of
fat in wet feces. These methods depend upon various principles and
generally give correct results. However, although they require less
time than methods in which feces are first dried, they are still quite
time-consuming.

The method described in this volume requires only 30–45 minutes.
It was elaborated from the principles published by von Liebermann
and Székely (1) and by Saxon (2).

Principle

Feces are saponified with concentrated potassium hydroxide in
ethanol, giving a solution which contains the soaps derived from
the neutral fats and the fatty acids and also the soaps which were
originally present in the stool. By adding HCl to the alkaline solution, the fatty acids are liberated. Ethanol is then added and the
fatty acids are extracted with petroleum ether. The concentration
of ethanol is so chosen that, after the mixture has been shaken, the

* Based on the method of van de Kamer, ten Bokkel Huinink, and Weijers (4).

34

petroleum ether and the acid ethanol layers separate quickly; this separation is expedited by adding a small amount of amyl alcohol. Separation is complete after 10 minutes. In an aliquot sample of the petroleum ether layer, the fatty acids are titrated with alkali; thymol blue is used as an indicator.

Reagents

1. *Ethanol, 96%, containing 0.4% amyl alcohol.*
2. *Ethanol, 96%, neutral to thymol blue.*
3. *KOH, 33%.*
4. *HCl, 25%, specific gravity 1.13.*
5. *Petroleum ether, b.p. 60°–80°C. or 40°–60°C.:* When evaporated to dryness, it must leave no residue which can be titrated or saponified with alkali.
6. *NaOH, 0.1000 N.*
7. *Thymol blue, 0.2% in 50% ethanol.*

Special Equipment

150-ml. Erlenmeyer flasks with wide mouths (high model), each provided with a reflux condenser.

25-ml. pipet, fitted into the flask as illustrated in Fig. 1. With this apparatus the petroleum ether solution is brought into the pipet by blowing in order to prevent evaporation.

Procedure

Weigh about 5 g. of feces in a 150-ml. Erlenmeyer flask and record the exact weight. Add 10 ml. of 33% alkali and 40 ml. of ethanol containing 0.4% amyl alcohol; boil the mixture for 20 minutes under a reflux condenser; cool thoroughly. Add 17 ml. of 25% HCl, measured in a graduated cylinder, and cool the mixture again. Add exactly 50.0 ml. of petroleum ether and close the flask with a glass stopper. Shake vigorously for 1 minute. Transfer, after separation, 25.0 ml. of the petroleum ether layer into a small Erlenmeyer flask by using the pressure pipet.

Add a piece of filter paper; evaporate the petroleum ether and add 10 ml. of neutral ethanol. Titrate the fatty acids with 0.1 N NaOH from a microburet, using thymol blue as an indicator, until the yellow color begins to change.

Calculation

The calculations are carried out assuming an average molecular weight of 265 for fatty acids in stool. Hence the fatty acids in grams per 100 g. feces can be calculated as:

$$\frac{A \times 265 \times 1.04 \times 2 \times 100}{10,000 \, Q} = 5.51 \frac{A}{Q}$$

A = milliliters of 0.1 N alkali used in titration.

Q = grams of feces taken for analysis.

The correction factor 1.04 must be used since the petroleum ether layer increases 1% in volume when shaken with alcoholic hydrochloric acid and because 3% of the amount of fatty acids remains in solution in the acid alcoholic layer. Corrections for evaporation of the petroleum ether layer and for volume increase resulting from solution of fatty acids in it may be neglected. That the factor used is correct was demonstrated by the fact that the same fat content is found when the solution is extracted quantitatively with petroleum ether and when an aliquot is used as described above.

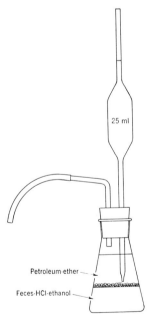

25 ml

Petroleum ether

Feces-HCl-ethanol

Fig. 1. Erlenmeyer flask with pressure pipet to remove aliquot of petroleum ether.

Precautions and Notes

In order to prevent irregular boiling during evaporation of the petroleum ether, a small piece of filter paper is used. No pumice must be used, as this absorbs fatty acids.

It is not necessary to evaporate the petroleum ether quantitatively, as small amounts of this solvent have no influence on the titration.

For the extraction of the acid alcoholic solution with petroleum ether, 60% alcohol appears to be most suitable. With lower concentrations, emulsions are easily formed, while at higher concentrations too much fatty acid remains dissolved in the alcoholic layer.

The extraction is complete after 1 minute; no more fat is extracted after shaking the mixture for 2, 3, 4, 5, or 10 minutes.

Usually phenolphthalein is used as indicator in fatty acid titration. In the yellow-colored petroleum ether solution, however, the change in color of thymol blue from yellow to green is much more evident than the change of phenolphthalein from yellow to reddish yellow. The two indicators have the same pH range, that is, from pH 8 to 10.

Titration of the fatty acid was preferred to weighing because it is quicker and because unsaponifiable matter is excluded at the same time.

The molecular weight of the fatty acids in stool depends on the kind of fat consumed; for general purposes, both in cases of steatorrhea and normal fat excretion, 265 was chosen—palmitic acid being the main component.

It appeared that fatty acids and fat added to feces could be recovered with a relative error of ±2%.

One reviewer (M. P.) found this method "far superior" to that of Saxon (2). Another (D. T.) found comparable results, in clinical studies, between the method using I^{131}-labeled fat and this method.

Application

The determination of total fatty acids in stool is carried out to study the fat retention in a patient by means of a fat balance.

The total amount of fat consumed in 24 hours, less the amount of excreted fat, represents the amount of retained fat. This may be expressed as per cent of the amount of fat consumed; the coefficient of fat retention is thus calculated.

A coefficient of fat retention can be calculated if the diet contains at least 50 g. of fat daily. If lower quantities are consumed, the fat in the feces originating from bacteria, intestinal cells, or bile (a total of 1–2 g. are excreted every 24 hours) will disturb the figures.

In order to decide if a steatorrhea exists or not, one has to pool the feces of at least 5 consecutive days. After determination of the fat content of this mixture, the mean coefficient of fat retention during this period can be calculated.

In metabolic investigations running during several weeks or even months, it is of advantage to calculate and note the sliding mean of 3 days in order to exclude, for the most part, the influence of daily fluctuations caused by the variability of intestinal motility (3).

In some exceptional cases it may be of value to determine un-split and split fat, i.e. fatty acids, separately. This is possible using the method described slightly modified (4).

Since most of the chemical methods for making a fat-balance study are too laborious to be practical for routine analysis, the coefficient of fat retention is often evaluated by microscopic exam-ination of the feces. This method, however, is absolutely unreliable, as can be demonstrated by comparing the results of microscopic determinations with those of chemical analysis (3). To evaluate the coefficient of fat retention by expressing the fat content of a ran-dom sample of feces as percentage of dry matter is equivocal (3).

Also the valuation of fat absorption by a chylomicron count or by interpreting a vitamin A tolerance curve is open to serious criti-cism (3). For example, it is probable that only those triglycerides that are absorbed via the lymphatics cause an increase in chylo-microns and vitamin A, whereas triglycerides (fatty acids) which are absorbed by the portal route do not give this increase (5, 6).

Normal Values

Since it is impossible to draw conclusions concerning steatorrhea from the fat content of feces, regardless of the amount of feces excreted in 24 hours, and the amount of fat consumed in the same period, only normal values for the coefficient of fat retention are given.

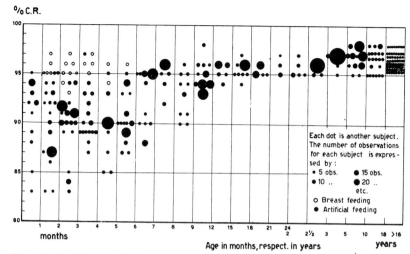

Fig. 2. Coefficients of fat retention in normal babies, toddlers, children, and adults. From Weijers and van de Kamer (3).

For healthy adults eating butter, margarine, or oils, a coefficient of fat retention of 95–98% is found. If sheep tallow is consumed, a coefficient of retention of approximately 90% will be found.

In babies, however, various coefficients can be found, as is shown in Fig. 2. Only in cases of breast feeding is the coefficient of fat retention as high as 95%.

REFERENCES

1. von Liebermann, L., and Székely, S., Eine neue Methode der Fettbestimmung in Futtermitteln, Fleisch, Koth, etc. *Arch. ges. Physiol. Pflügers* **72**, 360–366 (1898).
2. Saxon, G. J., A method for the determination of the total fats of undried feces and other moist masses. *J. Biol. Chem.* **17**, 99–102 (1914).
3. Weijers, H. A., and van de Kamer, J. H., Coeliac disease. I. Criticism of the various methods of investigation. *Acta Paediatr.* **42**, 24–33 (1953).
4. van de Kamer, J. H., ten Bokkel Huinink, H., and Weijers, H. A., Rapid method for determination of fat in feces. *J. Biol. Chem.* **177**, 347–355 (1949).
5. Bloom, B., Chaikoff, I. L., and Reinhardt, W. O., Intestinal lymph as pathway for transport of absorbed fatty acids of different chain lengths. *Am. J. Physiol.* **166**, 451–455 (1951).
6. Fernandes, J., van de Kamer, J. H., and Weijers, H. A., The absorption of fats studied in a child with chylothorax. *J. Clin. Invest.* **34**, 1026–1036 (1955).

GAMMA GLOBULIN IN SERUM*

Submitted by: Howard S. Friedman, 6580th USAF Hospital, Holloman Air Force Base, New Mexico†

Checked by: David Miller, McGuire Veterans Administration Hospital, Richmond, Virginia

Nelson F. Young, Medical College of Virginia, Richmond, Virginia

Introduction

The determination of serum gamma globulin is an important diagnostic aid which has been recognized chiefly since the application of the electrophoretic method to pathologic serums. It has been known for some time that gamma globulin was present in normal serum, but chemical methods for its separation and estimation have only recently appeared in the literature. Boundary electrophoresis is seldom considered essential in the small laboratory, but, although the more easily performed paper electrophoresis may eventually become widely used as a routine procedure for quantitative protein fractionation, there will always be a need for a reliable, simple procedure for the quantitative estimation of serum gamma globulin.

Early methods for the separation of gamma globulin from serum involved the use of a number of salts in varying concentrations. Among these was 33% saturated ammonium sulfate solution. The technique for using this solution has been modified and improved and forms the basis of the present quantitative method. Another type of procedure takes advantage of the formation of turbidity resulting from the combination of protein with heavy metals such as zinc, copper, cadmium, and barium under controlled conditions of temperature, pH, and concentration of protein and reagents. Wolfson and Cohn (1) developed a procedure for separating the serum proteins into four fractions which resemble electrophoretic

---

* Based on the method of Wolfson, Cohn, Calvary, and Ichiba (1).

† Present address: 7520th USAF Hospital, London, England.

40

fractions. The separation of gamma globulin is based upon its insolubility in 33% saturated ammonium sulfate. The pH is adjusted to 6.4 (the isoelectric point). Sodium chloride is added for maximum recovery.

Principle

Serum gamma globulin is precipitated from a solution containing 18.5 g. of ammonium sulfate and 2.9 g. of sodium chloride per 100 ml. This solution is 1.4 M (33% saturated) in ammonium sulfate and 0.5 M in sodium chloride. The protein is precipitated in the cold and the supernatant removed after centrifugation. The protein precipitate is dissolved in sodium chloride solution and the intensity of the color developed after the addition of biuret reagent is estimated photometrically.

Reagents

1. *Ammonium sulfate, 19.5% (in 2.93% sodium chloride).*

Note 1: The original concentration of ammonium sulfate must be 19.5%, since 1 volume of serum is added to 19 volumes of reagent. If other dilutions, such as 1:10 or 1:25 are used, the original concentrations must be adjusted accordingly, so that the final concentrations will be 18.5% and 2.9%, respectively.

To approximately 700 ml. of water in a 1-l. volumetric flask, add 195 g. of reagent grade ammonium sulfate and 29.3 g. of reagent grade sodium chloride. Dissolve and bring to a volume of about 995 ml. Mix well. Adjust the pH to 6.4 by the dropwise addition of concentrated ammonium hydroxide or sulfuric acid as needed.

Note 2: Although this step is not included in the original procedure, it is a desirable one, because it decreases the tendency of alpha and beta globulins to be precipitated.

Dilute to the mark with water and store in a screw-cap bottle, preferably of polyethylene.

2. *Sodium chloride, 0.9%.* Dissolve 9.0 g. of reagent grade sodium chloride in water and make to 1 l.

3. *Biuret reagent (stock).* Dissolve 45 g. of reagent grade sodium potassium tartrate (Rochelle salt) in about 400 ml. of 0.2 N NaOH in a 1-l. volumetric flask. Add 15 g. of $CuSO_4.5H_2O$, reagent-grade fine crystals, while stirring and continue to stir until dissolved. Add 5 g. of reagent-grade KI and dilute to 1 l. with 0.2 N NaOH.

Note 3: The biuret reagent described here has been selected to conform with that used in Volume I of this series. An alternate reagent (preferred by the submitter) is a modification of that of Gornall et al. (2). This reagent consists of 6 g. of sodium potassium tartrate, 1.5 g. of $CuSO_4.5H_2O$, and 300 ml. of 10% NaOH per liter. A tenfold concentration may be prepared. The addition of KI to this reagent has been found to be unnecessary. The stock reagent and the dilute reagent are both extremely stable.

4. Biuret reagent (dilute). Dilute 1 volume of stock biuret reagent with 9 volumes of 0.2 N NaOH containing 5 g. of KI per liter. (See note 3.)

5. Sodium hydroxide, (A) 0.2 N and (B) 0.2 N with 0.5% KI. (A) Prepare from carbonate-free concentrated sodium hydroxide solution (18 N). Titrate against standard acid. Add 5 g. of KI per liter to obtain (B).

6. Protein standard: Pooled normal or pathologic serums may be used. Standardize by the Kjeldahl method and correct for NPN. The standard bovine albumin prepared by Armour and Company and the Protein Standard marketed by Standard Scientific Supply Company are both suitable standards. Dilute the standard with 0.9% sodium chloride to prepare solutions containing 25, 50, 75, and 100 mg. of protein per 100 ml., which will be equivalent to 0.5, 1.0, 1.5, and 2.0 g. of gamma globulin per 100 ml. of serum, respectively.

Note 4: Prepare dilutions on the day of use and discard the excess. The standardized serum may be kept several months in the frozen state.

Procedure

In a 12-ml. thick-wall borosilicate conical centrifuge tube, place 5.7 ml. of ammonium sulfate reagent. With a 1-ml. serologic pipet, carefully overlay the reagent with 0.30 ml. of clear unhemolyzed serum.

Note 5: Plasma cannot be used in this procedure unless correction is made for the fibrinogen which is also precipitated by the reagent.

If a high gamma globulin is expected, as indicated by previous results or otherwise, one-half or one-third quantities may be used at this point. Multiply the results by 2 or 3, respectively. (See also Note 10.)

As little as 0.10 ml. may be used for this separation. Increase the amount of reagent accordingly, so that the total volume of precipitating mixture is 6.0 ml.

Close the tube with the thumb protected by a rubber finger cot and gently invert the tube several times to a visible maximum turbidity. Place the tube in an ice bath for 15 minutes. Centrifuge the tube at 3000 r.p.m. for 10 minutes. Remove the supernatant by aspiration and return the tube to the centrifuge for 5 more minutes. Remove any additional supernatant by aspiration.

NOTE 6: Virtually all of the supernatant must be removed because the ammonia-copper complex color may interfere in the estimation of color intensity.

Dry the inside of the tube with a rolled filter paper or a dry pad of gauze. Add 2.00 ml. of 0.9% sodium chloride solution and agitate gently until the precipitate dissolves. Stir with an applicator stick if necessary.

NOTE 7: If the volume of the precipitate is large (greater than 0.3 ml.), dissolve it in 4.00, 6.00, or more milliliters of 0.9% sodium chloride. Transfer 2.00 ml. of this solution to a cuvette and add 5.00 ml. of dilute biuret reagent. Multiply the result by the appropriate factor.

Place 2.00 ml. of 0.9% sodium chloride solution in a 19 × 150 mm. test-tube cuvette, for the blank. To both tubes add 5.00 ml. of dilute biuret reagent and mix. Let stand at room temperature for 10 minutes. Pour the unknown into a second cuvet and read in the spectrophotometer, setting the absorbance of the blank to 0, at 555 mµ.

NOTE 8: If the absorbance of the unknown is higher than that of the highest calibration standard, add 2 ml. of 0.9% sodium chloride solution and 5 ml. of dilute biuret reagent to the cuvet, mix well, and read again. Multiply the calculated results by 2.

Determine the protein concentration by referring to a calibration chart or curve prepared by treating 2.00 ml. of each of the dilute standards with 5.00 ml. of dilute biuret reagent under the conditions stated above. Divide this result by 3.

NOTE 9: The calibration standards described in relation to the protein standard (reagent No. 6) represent 1:20 dilutions of serum, or 0.1 ml. of serum or the precipitate therefrom in a final volume (before the addition of dilute biuret reagent) of 2.0 ml. Since the procedure requires 3 times this amount of serum, it is necessary to divide the results from the calibration chart or curve by 3. If different quantities of serum or different final volumes of solution are used, appropriate factors must be used in the calculations.

In the range from 0.5 to 1.5 g. of gamma globulin per 100 ml. of serum, the absolute error is less than 0.1 g./100 ml. The relative error is 6–20%.

In the range from 1.5 to 4.5 g. of gamma globulin per 100 ml. of serum, the absolute error is about 0.1 g./100 ml. The relative error is approximately 2–6%.

Range of Values

The range in the submitter's laboratory is 0.65 g. to 1.70 g. per 100 ml. with 80% of the values below 1.50. The average value is 1.28, with a standard deviation of 0.28 in a series of 30 normal serums. The range by electrophoretic techniques is 0.61 to 0.91 g. per 100 ml.

In general, low values are associated with wasting and neoplastic processes, while high values are found in infectious and inflammatory diseases. Agammaglobulinemia is a congenital or acquired state in which the body loses its ability to produce antibodies, which make up the bulk of the gamma globulin fraction; hence, little or no gamma globulin is found in serum of persons with agammaglobulinemia.

Discussion

It has been established by the work of E. J. Cohn and his associates (3) that the electrophoretic separation of six distinct protein fractions in plasma and serum is far from complete. These workers have separated serum proteins into at least 30 fractions on the basis of chemical, physical, and physiologic properties, by taking advantage of ion-protein and protein-protein interactions at various metalion concentrations, at low temperatures, and with controlled pH in the presence of dilute ethanol. In their scheme (3), the gamma globulins are included in the group II fraction and include the antibodies to many foreign proteins.

Owing to the size, shape, molecular weight, and other physical and chemical characteristics of gamma globulin, it is difficult to separate it from closely related compounds such as alpha and beta globulins. There is a great tendency to occlusion.

In the salting out procedures, serums from healthy persons show values which are very similar to each other and quite reproducible from an analytical point of view. In disease states, however, the synthesis of proteins may vary from the normal in two ways. Abnormal quantities of a fraction may be produced, in which case there is a *dysproteinemia*, or abnormal types of protein may be produced, and this may be termed *paraproteinemia*. In the former type, fractionation will determine the abnormal increase correctly, since the increase will be due to proteins similar in kind to those already present in normal serum. In the latter case, the protein may be quite different from the normal compounds, varying in its mobility, shape, size, molecular weight, isoelectric point, or other characteristics. It may therefore appear in electrophoresis as an increase within a given fraction, or it may appear between two fractions. Unfortunately, the methods for determining which is the case are extremely difficult and time-consuming and generally require highly specialized apparatus and technical ability. Thus, electrophoretic and chemical fractionation procedures may be subject to the same error. This is illustrated most strikingly in the case of multiple myeloma, of which four types may appear. The first three show increases in globulin in the alpha, beta, or gamma fraction. The fourth type is a paraproteinemia showing the production of an abnormal protein called "M" protein, whose electrophoretic properties cause it to appear between the beta and gamma fractions. When a serum containing this "M" protein is fractionated by precipitation, the abnormal protein may appear in the beta fraction, or it may appear in the gamma fraction, or in both, depending upon its similarity in physical properties to beta or gamma globulin. Paraproteinemias may occur occasionally in other disease conditions. This may explain some of the discrepancies often found between electrophoretic and chemical protein analyses.

Wolfson *et al.* (1) found that the reagent described above gave about 95% recovery of gamma globulin added to serum. They further noted that all of the gamma globulin was precipitated when pure solutions were analyzed.

The two checking laboratories undertook a comparison of electrophoretic data with results obtained by the chemical procedure described.

Samples of blood were obtained and centrifuged. Chemical analysis was performed on the same or the following day; electrophoresis was performed as promptly as possible with samples of serum which were frozen when a delay of more than 2 days was unavoidable. All chemical analyses were performed in one laboratory (D. M.). The electrophoretic studies were performed in the other (N. F. Y.) without knowledge of results of chemical analyses.

Electrophoresis was performed in Veronal buffer pH 8.6, ionic strength 0.1. The Klett apparatus was used routinely with the

TABLE I

COMPARISON OF CHEMICAL AND ELECTROPHORETIC DETERMINATIONS OF SERUM
GAMMA GLOBULIN

| Serum no. | Total protein | Gamma-globulin | |
		Chemical procedure	Electrophoretic procedure
	%	%	%
1	6.2 g.	2.2 g.	2.1 g.
2	7.4	2.9	2.9
3	6.8	1.5	1.3
4	6.1	2.3	2.5
5	6.6	1.2	0.8
6	3.5	Turbid only	0.2
7	7.9	1.9	1.4
8	6.8	1.3	1.2
9	5.9	2.2	1.2
10	7.9	2.9	3.0
11	6.1	1.5	1.3
12	5.5	2.7	2.4
13	5.9	2.2	2.0
14	6.1	1.2	1.2
15	9.7	3.0	5.5

11-ml. cell for 180 minutes with a current of approximately 12 milliamperes and 138 volts. The enlarged Longsworth patterns were traced and subjected to planimetry, and the percentage of gamma globulin was estimated.

These results were compared with the ammonium sulfate fraction described here. The same protein standard was used for both chemical gamma globulin and total protein determinations. A total of 15 serums were compared with the results shown in Table I.

The results in 12 instances are acceptable without comment. Serum sample No. 5 is unusual in that it had an electrophoretic A:G ratio of more than 2:1. Serum No. 9, from a patient suffering from malnutrition, showed a preponderance of both beta and gamma globulins. The discrepancy may be attributable to an increase of beta globulin, which was probably of such physical properties as to be precipitated as gamma globulin. Serum No. 15 was from a patient with myeloma of the gamma type. This abnormal protein was presumably not completely precipitated.

The biuret reaction for the measurement of proteins is based upon the color developed by the peptide bond —CO—NH— in strongly alkaline solutions containing small concentrations of copper ions. The color developed per unit weight of protein depends upon the number of peptide bonds in the solution and not upon the amino acid composition. A tenfold increase in molecular weight of a protein would result in an increase of less than 0.5% in the amount of peptide bond. This residue, therefore, is most constant in relative value from one protein to another.

The biuret reagent develops color with a number of compounds other than proteins; these include amino acids, peptides, biuret, oxalamide, and ammonium ion. The latter is the only substance that might interfere in the present instance, since the amino acids present in serum are in very low concentration. The submitter (H. S. F.) has found that an amount of the precipitating reagent equal to 0.5 ml. of precipitate will give an equivalent concentration of gamma globulin of less than 0.17 g./100 ml. The specific extinction coefficient in this method is 42, and the molecular extinction coefficient for gamma globulin, assuming a molecular weight of 160,000, is about 6,700,000.

Other Methods

Serum gamma globulin has been determined turbidimetrically with ammonium sulfate (4–7), by immunochemical means (8), by sodium sulfate precipitation (9), and by zinc sulfate (10). In general, the serum gamma globulin fraction is precipitated by 13.5% sodium sulfate (9), 18.5% ammonium sulfate (1, 4–7), and 20% ethanol (11).

REFERENCES

1. Wolfson, W. Q., Cohn, C., Calvary, E., and Ichiba, F., Studies in serum proteins. V. A rapid procedure for the estimation of total protein, true albumin, alpha globulin, beta globulin and gamma globulin in 1.0 ml. of serum. *Am. J. Clin. Pathol.* **18,** 723–730 (1948).
2. Gornall, A. G., Bardawill, C. J., and David, M. M., Determination of serum proteins by means of the biuret reaction. *J. Biol. Chem.* **177,** 751–766 (1949).
3. Cohn, E. J., Strong, L. E., Hughes, W. L., Jr., Mulford, D. J., Ashworth, J. N., Melin, N., and Taylor, H. L., Preparations and properties of serum and plasma proteins. IV. A system for the separation into fractions of the proteins and lipoprotein components of biological tissues and fluids. *J. Am. Chem. Soc.* **68,** 459–475 (1946).

4. De la Huerga, J., and Popper, H., Estimation of serum gamma-globulin concentration by turbidity. *J. Lab. Clin. Med.* **35,** 459–465 (1950).
5. Jager, B. V., Schwartz, T. B., Smith, F. L., Nickerson, M., and Brown, D. M., Comparative electrophoretic and chemical estimations of human serum albumin; an evaluation of six methods. *J. Lab. Clin. Med.* **35,** 76–86 (1950).
6. Kunkel, H. G., Estimation of alterations of serum gamma-globulin by turbidimetric technique. *Proc. Soc. Exptl. Biol. Med.* **66,** 217–224 (1947).
7. Looney, J. M., and Amdur, M. O., Determination of gamma-globulin in blood serum. *Federation Proc.* **8,** 220 (1949).
8. Kabat, E. A., Glusman, M., and Knaub, V., Quantitative estimation of albumin and gamma-globulin in normal and pathological cerebrospinal fluid by immunochemical methods. *Am. J. Med.* **4,** 653–662 (1948).
9. Kibrick, A. C., and Blonstein, M., Fractionation of serum into albumin and alpha, beta and gamma-globulins by sodium sulfate. *J. Biol. Chem.* **176,** 983–987 (1948).
10. Kunkel, H. G., Ahrens, E. H., Jr., and Eisenmenger, W. J., Application of turbidimetric methods for estimation of gamma globulin and total lipid to the study of patients with liver disease. *Gastroenterology* **11,** 499–507 (1948).
11. Cohn, E. J., The properties and functions of the plasma proteins, with a consideration of the methods for the separation and purification. *Chem. Revs.* **28,** 395–417 (1941).

HEMOGLOBIN*

Submitted by: ADRIAN HAINLINE, JR., Department of Clinical Pathology, The Cleveland Clinic Foundation, and The Frank E. Bunts Educational Institute, Cleveland, Ohio

Checked by: J. WAIDE PRICE, School of Medicine, Western Reserve University, Cleveland, Ohio

S. P. GOTTFRIED, Bridgeport Hospital, Bridgeport, Connecticut

Introduction

The estimation of hemoglobin is one of the most widely used tests performed by clinical laboratories. Yet with regard to accuracy the results of hemoglobinometry are poor compared with results of other determinations. This inaccuracy is due to the diversity of methodology and to a greater extent to the lack of an easy means of standardization. The methods for hemoglobin range from very simple to rather complicated techniques. Complexity of the method does not necessarily correspond with accuracy; in fact, the simplest procedure of all is among the most accurate.

The difficulty of isolating hemoglobin in the pure and uniform state (1) makes it necessary to standardize a hemoglobin method with a second method. The indirect methods which can be used for hemoglobin standardization are too lengthy and difficult to be of use for routine analysis. Procedures most widely used include iron analysis and oxygen capacity of hemoglobin solutions. Human hemoglobin is 0.34% iron (2) or will absorb 1.34 volumes per cent of oxygen per gram (3). Oxygen-capacity methods require a special apparatus and considerable experience with the determination to give accurate results. The determination of iron, on the other hand, only requires equipment that should be present in any clinical chemistry laboratory.

Until recently, attempts to introduce a uniform standard have been disorganized and ineffective. In England, the use of a stand-

* Based on the methods of Szigeti (10), Drabkin and Austin (12), and Wong (14) as modified by Hainline.

ardized visual colorimeter has been a satisfactory answer to the problem (4), although for many years the standard in use was incorrect. During that period the hemoglobin levels reported in England were low; nevertheless there was good agreement between the laboratories throughout the country. In the United States, Sunderman and others (5, 6) proposed that the iron determination be used by each clinical laboratory to standardize its hemoglobinometry. Shortly afterward a standardization committee* (7) proposed a standard that could be distributed among the clinical laboratories. This standard was to be a stable solution of cyanmethemoglobin which would be prepared from hemoglobin standardized according to its iron content. This standard has the advantage that the same solutions used in preparing it can be used for routine analysis.

As a routine procedure for hemoglobin determination, the oxyhemoglobin method has much to offer in the way of stability, speed and accuracy. The first spectrophotometric method for oxyhemoglobin was reported by Davis and Sheard (8) in 1927 and was used by Sheard and Sanford in their first practical photoelectric hemoglobinometer (9). These authors used 0.1% sodium carbonate as a diluent for blood. As there is some tendency for fading in the dilute sodium carbonate, other slightly alkaline solutions have been suggested for diluents. The best of these, approximately 0.007 N ammonium hydroxide, was suggested by Szigeti (10).

Two methods which are suitable for routine use will be discussed: the oxyhemoglobin method and the cyanmethemoglobin method. The third method for total iron is suitable for standardization of either of the other two procedures. Actually, once standardized either cyanmethemoglobin or oxyhemoglobin can be used to standardize the other method. Because of its greater stability, the cyanmethemoglobin is best suited for standardization.

I. OXYHEMOGLOBIN METHOD

Principle

Whole blood is diluted with a very weak ammonia solution and oxygenated by exposure to atmospheric oxygen. The intense color of oxyhemoglobin in the weakly alkaline solution is measured photo-

* *Ad hoc* Panel on the Establishment of a Hemoglobin Standard of the Division of Medical Sciences, National Academy of Sciences, National Research Council.

metrically. The procedure is standardized with blood which has been analyzed for iron.

Reagents

Ammonium hydroxide, approximately 0.007 N. Add 4 ml. of commercial reagent-grade NH_4OH to 1 l. of water and mix. It is best to keep this solution in a borosilicate glass bottle. It may be necessary to filter solutions that have been standing for some time if any turbidity develops.

Procedure

Measure 5.0 ml. of 0.007 N ammonium hydroxide into a photometer cuvette. (If conditions require a larger volume, 10.0 ml. of ammonium hydroxide is equally satisfactory.) Carefully transfer 0.020 ml. of well-mixed blood with a Sahli-type hemoglobin pipet to the ammonium hydroxide solution. Rinse the pipet three times with the diluting fluid, stopper the cuvette with a clean rubber stopper, and shake several times to mix and oxygenate the hemoglobin. Read in a photometer using a green filter (540 mμ) against a 0.007 N ammonium hydroxide blank.

NOTE: Readings can be made within a few seconds or, if the cuvette is kept stoppered, at any time up to 2 or 3 days.

Calculation

The results are estimated by converting photometric values to grams of hemoglobin per 100 ml. from a calibration curve. Blood that has been analyzed for total iron is used as a standard to prepare the calibration curve.

Standardization of the Oxyhemoglobin Method

The calibration curve for the oxyhemoglobin method is prepared by obtaining several fresh specimens of oxalated normal blood, combining them, and analyzing the pooled blood for total iron or by making individual total iron determinations on several specimens. The blood sample for the iron determination should be measured out at the same time that the oxyhemoglobin calibration curve is begun. Transfer 0.100 ml. of well-mixed blood to 25.0 ml. of 0.007 N ammonium hydroxide; mix by inversion; shake for 10 seconds; and read in a photometer using a green filter (540 mμ). One sample of blood may be used to obtain a range of concentrations lower than the original concentration by making suitable dilutions with the ammonium hydroxide solution. For example, if 5.0 ml. of the oxyhemoglobin solution is diluted to 10.0 ml. and is mixed, the hemoglobin concentration of this solution can be considered half that of the original. If a more concentrated blood is desired, it may be necessary to remove from 10 to 20% of the plasma after centrifugation and to resuspend the cells in the remaining plasma. If transmittance values are plotted against hemoglobin concentrations on semilogarithmic graph paper, a straight line should result. Absorbance plotted against concentration on rectangular coordinate paper is also suitable.

II. CYANMETHEMOGLOBIN METHOD

Principle

Whole blood is diluted with a potassium ferricyanide-cyanide solution which oxidizes hemoglobin to methemoglobin, which is in turn converted to cyanmethemoglobin. The intensity of the cyanmethemoglobin color is measured photometrically. The procedure is standardized with blood which has been analyzed for iron.

Reagents

Potassium ferricyanide-cyanide solution. Transfer to a 1-l. volumetric flask the following reagent-grade chemicals: 1 g. $NaHCO_3$, 50 mg. KCN, and 200 mg. $K_3Fe(CN)_6$. Dilute with water to 1 l. No more than a month's supply should be made at one time because the solution slowly deteriorates. The solution should be stored in a brown bottle and kept out of direct light.

Procedure

Measure 5.0 ml. of the mixed cyanide solution into a photometer cuvette. Carefully transfer 0.020 ml. of well-mixed blood with a Sahli-type hemoglobin pipet to the cyanide solution. Rinse the pipet three times with the diluting fluid, stopper the cuvette with a clean rubber stopper, and invert several times. Allow the tube to stand for at least 10 minutes. Read in a photometer, using a green filter (540 mμ) and use mixed cyanide solution as a blank.

Calculation

The results are estimated by converting photometric values to grams of hemoglobin per 100 ml. from a calibration curve. Blood that has been analyzed for total iron is used as a standard to prepare the calibration curve.

Standardization of the Cyanmethemoglobin Method

The calibration curve for the cyanmethemoglobin method is prepared by analyzing pooled, fresh, oxalated, human blood for total iron or by doing individual total-iron determinations on several fresh specimens. The blood sample for iron determination should be measured out at the same time that the cyanmethemoglobin calibration curve is begun. Transfer 0.100 ml. of well-mixed blood to 25.0 ml. of mixed cyanide; mix by inversion; shake for 10 seconds; stand for 10 minutes; then read in a photoelectric colorimeter with a green filter (540 mμ). One sample of blood may be used to obtain a range of concentrations lower than the original concentration by making suitable dilutions with the mixed cyanide solution. For example, if 5.0 ml. of the cyanmethemoglobin solution is diluted to 10.0 ml. and is mixed, the hemoglobin concentration of this solution can be considered half that of the original. If a more concentrated blood is desired, it may be necessary to remove from 10 to 20% of the plasma after centrifugation and to resuspend the cells in the remaining plasma. Cyanmethemoglobin solutions follow Beer's law over the range of the method.

NOTE: For those who possess an accurate spectrophotometer, the spectrophotometric method described for oxygen capacity by Hickam and Frayser (11) is a very simple procedure: In each of two 100-ml. volumetric flasks place approximately 75 ml. of water. Label one as a blank; to the other add 0.5 ml. of well-mixed blood. To both flasks add 2 ml. of 4% K$_3$Fe(CN)$_6$. Wait 20 minutes. Add to each flask 0.5 ml. of 5% KCN and water to volume. Read the absorbance with a narrow slit opening at 540 mμ. Grams of hemoglobin = $A \times 29.0$. [Calculation based on millimolar extinction coefficient for cyanmethemoglobin of 11.5; see reference (12)].

III. TOTAL-IRON METHOD

Principle

Whole blood is digested with a 1:1 mixture of sulfuric and nitric acids to remove organic materials and to oxidize the iron to the ferric state. The diluted digest is treated with potassium persulfate and potassium thiocyanate to produce a color which can be read in a photometer. The concentration of hemoglobin can be calculated from the iron content.

Reagents

1. *Water, iron-free.* If necessary, distill demineralized water or ordinary distilled water in an all-glass still. To test for the presence of iron, add to 10 ml. of water, 1 ml. iron-free concentrated hydrochloric acid, 0.1 ml. concentrated nitric acid, 1 ml. 3 N potassium thiocyanate, and 2 ml. amyl alcohol. Shake thoroughly. The alcohol should remain colorless. Small laboratory deionizers may provide a satisfactory substitute for redistillation.

NOTE: All glassware must be iron-free. Glassware to be used for iron determinations should be rinsed three times with iron-free water in addition to rinses with ordinary distilled water. If the glassware is contaminated with iron, it may be necessary to rinse it with warm dilute nitric acid before the distilled water rinsings.

2. *Potassium persulfate, saturated.* Shake 7 g. reagent-grade $K_2S_2O_8$ with 100 ml. of iron-free water in a colored-glass bottle. Permit the undissolved crystals to remain at the bottom of the container. Store in a refrigerator.

3. *Potassium thiocyanate, approximately 3 N.* Dissolve 146 g. of reagent-grade KSCN in about 300 ml. of iron-free water. Add 20 ml. of pure acetone and dilute to 500 ml. Mix, allow to stand overnight, and filter through tightly packed acid-washed glass wool if a residue remains. Discard when any coloration appears.

4. *Sulfuric acid, concentrated.* Use reagent-grade 96–98% H_2SO_4 .

5. *Nitric acid, concentrated.* Use reagent-grade 70% HNO_3 .

6. *Standard iron solution.* Weigh 100.0 mg. pieces of clean shiny iron wire, analyzed grade. Previous determination of the weight per unit length will aid in estimating the proper length to cut. Drop the wire into a boiling solution of dilute nitric acid (10 ml. of concentrated nitric acid and 40 ml. of iron-free water). Boil until completely dissolved, cool, and transfer quantitatively to a 100-ml. volumetric flask. Dilute to volume with iron-free water and mix. This solution is stable indefinitely. This stock standard contains 1.00 mg. of iron per milliliter.

NOTE: It is necessary to use hot or boiling nitric acid to assure that all of the iron is oxidized to the ferric state. Thiocyanate gives color only with ferric iron.

A dilute iron standard should be prepared on the day of use by diluting 10.0 ml. of stock standard to 100.0 ml. with iron-free water. This standard contains 0.100 mg. of iron per milliliter. From this solution, prepare for calibration dilute standards which contain 0.0020, 0.0040, and 0.0060 mg. of iron per milliliter. In the following procedure these concentrations correspond to bloods which contain 20, 40, and 60 mg. of total iron per milliliter. At least three concentrations should be used.

Procedure

Add 5.00 ml. of well-mixed whole blood to approximately 25 ml. of iron-free water in a 50-ml. volumetric flask. Mix well by swirling, dilute to mark, and mix. Best results will be obtained if the following analysis of the diluted blood is performed in triplicate. Transfer 1.00 ml. of the diluted blood to a borosilicate digestion tube calibrated at 25.0 ml.

NOTE: A borosilicate Lewis-Benedict tube (20 × 200 mm.) is recommended. Whatever the tube may be, the calibration should be rechecked. Digestion also may be carried out in an uncalibrated tube or micro-Kjeldahl flask and the blood cooled, diluted, and transferred quantitatively with washing to a 25-ml. volumetric flask.

To the diluted blood, add 1.0 ml. of concentrated sulfuric acid, 1.0 ml. of concentrated nitric acid, and one or two glass beads. Heat over a microburner until the copious fumes of sulfur trioxide fill the flask.

NOTE: If a hood is not available, a satisfactory substitute can be made from a funnel and a water vacuum pump.

The heating should proceed at a rate that will not drive off the nitric acid too rapidly at first. Digestion should require about 5 to 10 minutes.

NOTE: If the nitric acid is boiled off too rapidly, charring will occur when the sulfuric acid begins to decompose. If the digestion mixture is allowed to cool for a minute, 2 drops of 70% perchloric acid plus a minute more of heat will clear the mixture. However, the use of perchloric acid or hydrogen peroxide at this point may give low results.

At the completion of digestion the slight amber color which may remain will not interfere with the determination. Cool, add 10 ml. of iron-free water and mix. Cool to room temperature. Arrange a series of 25-ml. volumetric flasks for the blank and the standards.

NOTE: It is important that several checks be made on the reagent blank. All reagents, including the digesting acids, should be included. It is necessary to carry out the digestion step with the blank, as excess nitric acid will give color with the thiocyanate and increase the blank.

Into the first flask, place 10.0 ml. of iron-free water; into the remaining flasks place 10.0 ml. each of as many standards as are to be employed. To each of these and the unknown tubes add 0.10 ml. concentrated nitric acid, 1.0 ml. of concentrated sulfuric acid, 1.0 ml. of saturated potassium persulfate, and 2.0 ml. of potassium thiocyanate; dilute to volume and mix. Read in a photometer at 480 mμ. The transmittances of the standards are plotted against their concentrations on semilogarithmic graph paper. If the iron concentrations are expressed in terms of hemoglobin in grams per 100 ml., the hemoglobin concentration of the blood may be read directly from the graph. The iron concentration of human hemoglobin is 0.340% or 3.40 mg. of iron per gram of hemoglobin (1). To convert iron to hemoglobin, divide the number of milligrams of iron in 100 ml. of specimen by the number of milligrams of iron per gram of hemoglobin:

$$\text{Hemoglobin (grams/100 ml. of blood)} = \frac{\text{Milligrams of iron/100 ml.}}{3.40}$$

Discussion

Despite the existence of many simple methods, the estimation of hemoglobin is lacking in accuracy and uniformity in laboratories throughout the world. In a survey of clinical laboratories, Belk and Sunderman (13) found that only 37% of the analyses of two hemoglobin samples were within an acceptable range of ±3%. In more than 10% of the analyses, the error exceeded 1.5 g. of hemoglobin. Discrepancies between laboratories are due primarily to the lack of a primary hemoglobin standard. In addition, errors in measurements, use of unsuitable and obsolete methods, and differences in methodology contribute to the disagreements (5).

Because of its accuracy, the estimation of total iron in a specimen of blood is believed to be the best basis for determining its hemoglobin content. Hemoglobin contains 98% of the iron in blood (5). For clinical purposes, the nonhemoglobin iron usually is ignored, but for exact work it should be considered. To obtain a more representative sample of hemoglobins, several normal specimens of blood are pooled or it may also be satisfactory to analyze a number of specimens separately for iron and to use their values to obtain individual points on the calibration curve.

NOTE: A specimen taken from one individual conceivably could contain an abnormal hemoglobin, the iron content of which might be different from the normal.

In the standardization of a blood for hemoglobin content by the iron method, all determinations should be done at least in triplicate. The colorimetric procedures that use thiocyanate seem most adaptable to the usual laboratory. The procedure of Wong (14) for iron determination is most widely used and certain modifications are highly recommended (5). If the modifications in time suggested by Ponder (15) are necessary, digestion with nitric and sulfuric acid or with these acids plus perchloric acid (16) is simpler and requires fewer reagents and less time. Hot digestion can be accomplished in approximately 5 minutes. In the cold digestion of the Wong procedure, 5 minutes is required for the addition and mixing of just the first reagent—sulfuric acid—with the sample. In addition, considerable experience seems necessary to obtain reproducible results with the Wong procedure.

Oxygen-capacity methods are suitable for standardization, but unless already performed in the laboratory for other purposes, they are time consuming for preliminary preparation and may be inaccurate in the hands of inexperienced personnel. Values based on oxygen-capacity determinations tend to be slightly lower than those based on the iron content. Methemoglobin and carboxyhemoglobin contribute to these differences.

Errors in the oxyhemoglobin technique may be personal or due to faulty equipment. The greatest source of error in the determination is perhaps due to the small capacity of the pipet used to measure blood. Accurate pipet calibration is difficult, and errors in volume are frequent. If consistently accurate results are to be expected, routine checking of the calibration of new pipets should be done by the laboratory (5, 17).

NOTE: Correct calibration of volumetric glassware is of utmost importance for standardization by iron analysis. Note the relatively large amount of blood used as a sample in this procedure. This quantity of 5 ml. was chosen to reduce the error in pipetting small amounts.

The correct use of the pipet also is necessary. For example, caution must be observed not to rinse with diluent above the mark and to remove any blood adhering to the outside of the pipet. The measured specimen should be placed directly into the diluent and carefully rinsed from the pipet without spattering from excessive blowing.

Before the sample is removed from a specimen tube, the blood should be thoroughly mixed by inversion. Otherwise, the rapid sedimentation of cells may lead to low, nonreproducible results. If capillary blood is the source of the sample, the wound should be large enough to bleed freely and "milking" of the member should not be necessary. If pressure is applied around the wound, tissue fluids will be forced into the blood with subsequent dilution of the specimen. In general, all precautions that are necessary in the accurate determination of the hematocrit and the red blood-cell count should be observed in the hemoglobin determination.

If possible, the determination of hemoglobin in the laboratory should be accompanied by the hematocrit. These two analyses serve as a check upon each other.

The oxyhemoglobin technique will fail to give accurate indication of the hemoglobin level when the blood is excessively lipemic. It may be in error occasionally in a pronounced leukemia. Many of the colorimetric procedures, including the measurement by cyanmethemoglobin, also are affected by these conditions. If hemoglobin derivatives such as methemoglobin are present, the oxyhemoglobin method will not give as accurate results. The cyanmethemoglobin method (12) includes the measurement of the abnormal hemoglobin compounds and on that basis is the most accurate of the colorimetric methods. One disadvantage of the cyanmethemoglobin method is the use of cyanide as a reagent. However, it should be pointed out that the total amount of cyanide in 1 l. of the diluting fluid, recommended here, contains approximately one-fourth of the lethal dose (7), which certainly reduces the dangers encountered in careless pipetting. Nevertheless, it is a toxic solution. Care should be given to flush sinks well before and after discarding this solution to prevent any formation of HCN due to waste acids, such as carbon dioxide in the air, coming in contact with the cyanide. Considering the simplicity and accuracy of the oxyhemoglobin method, it seems unnecessary to expose the laboratory personnel to cyanide.

The oxyhemoglobin procedure as described has several advantages over the cyanmethemoglobin method. First, the dilute ammonium hydroxide requires a simple volumetric dilution of one reagent. The cyanide solution requires a balance, and three materials must be weighed. Secondly, and more important, to the user of large quantities of diluent, the cyanide solution is not stable much longer than a month, whereas the dilute ammonium hydroxide will last at least several months. Thirdly, the cyanmethemoglobin forms in 10 minutes, necessitating a wait, while the oxyhemoglobin can be used within a few seconds. Although cyanmethemoglobin is stable over a period of months, oxyhemoglobin is stable only a few days. For routine use the longer stability gives no advantage; as an interlaboratory standard, however, the cyanmethemoglobin has much in its favor because of its stability.

Range of Values

Hemoglobin concentration is usually high at birth but falls to a low level in a short time. It rises with increasing age. In boys and girls hemoglobin concentrations are equal up to the time of puberty. After puberty, the hemoglobin in the male continues to rise until age 16 or 17; the hemoglobin in the female tends to reach a plateau which is lower than the male level at its maximum. In pregnancy, the hemoglobin level generally falls but rises again after delivery.

The normal ranges that usually are accepted for clinical purposes are: men, 12.5–17.5 g. per 100 ml.; women, 11.5–15.5 g. per 100 ml. Because the range in healthy adults is so wide, no single value can or should be chosen as a 100% value.

REFERENCES

1. Sunderman, F. W., Copeland, B. E., MacFate, R. P., Martens, V. E., Naumann, H. N., and Stevenson, G. F., Hemoglobin standardizations. A commentary on procedures to insure reliable hemoglobinometry. *Am. J. Clin. Pathol.* **25,** 489–493 (1955).
2. Bernhart, F. W., and Skeggs, L., The iron content of crystalline human hemoglobin. *J. Biol. Chem.* **147,** 19–22 (1943).
3. Hüfner, G., Neue Versuche zur Bestimmung der Sauerstoffcapacität des Blutfarbstoffs. *Arch. f. Physiol.*, 130–176 (1894).
4. MacFarlane, R. G., and Poole, J. C. F., Haemoglobinometry and photoelectric erythrocyte counts. *Am. J. Clin. Pathol.* **24,** 67–70 (1954).
5. Sunderman, F. W., MacFate, R. P., MacFadyen, D. A., Stevenson, G. F.,

and Copeland, B. E., Symposium on clinical hemoglobinometry. *Am. J. Clin. Pathol.* **23,** 519–598 (1953).

6. Sunderman, F. W., Copeland, B. E., MacFate, R. P., Martens, V. E., Naumann, H. N., and Stevenson, G. F., Manual of workshop in clinical hemoglobinometry of American Society of Clinical Pathologists (condensed version). *Am. J. Clin. Pathol.* **25,** 695–713 (1955).

7. Cannon, R. K., Proposal for the distribution of a certified standard for use in hemoglobinometry. *Clin. Chem.* **1,** 151–156 (1955).

8. Davis, G. E., and Sheard, C., The spectrophotometric determination of hemoglobin. *A.M.A. Arch. Internal Med.* **40,** 226–236 (1927).

9. Sheard, C., and Sanford, A. H., A photo-electric hemoglobinometer. Clinical applications of the principles of photo-electric photometry to the measurement of hemoglobin. *J. Lab. Clin. Med.* **14,** 558–574 (1929).

10. Szigeti, B., Estimation of oxyhaemoglobin and of methaemoglobin by a photoelectric method. *Biochem. J.* **34,** 1460–1463 (1940).

11. Hickam, J. B., and Frayser, R., Spectrophotometric determination of blood oxygen. *J. Biol. Chem.* **180,** 457–465 (1949).

12. Drabkin, D. L., and Austin, J. H., Spectrophotometric Studies. II. Preparation from washed blood cells; nitric oxide hemoglobin and sulfhemoglobin. *J. Biol. Chem.* **112,** 51–65 (1935).

13. Belk, W. P., and Sunderman, F. W., A survey of the accuracy of chemical analyses in clinical laboratories. *Am. J. Clin. Pathol.* **17,** 853–861 (1947.)

14. Wong, S. Y., Colorimetric determination of iron and hemoglobin in blood. *J. Biol. Chem.* **77,** 409–412 (1928).

15. Ponder, E., Errors affecting the acid and the alkali hematin methods of determining hemoglobin. *J. Biol. Chem.* **144,** 339–342 (1942).

16. Dupray, M., A colorimetric method for the determination of iron and hemoglobin in the blood. *J. Lab. Clin. Med.* **12,** 917–920 (1927).

17. Ellerbrook, L. D., A simple colorimetric method for calibration of pipets. *Am. J. Clin. Pathol.* **24,** 868–874 (1954).

FREE AND CONJUGATED 17-HYDROXYCORTICOSTER-
OIDS IN PLASMA*

Submitted by: ALFRED M. BONGIOVANNI, Department of Endocrinology, Children's
Hospital of Philadelphia, University of Pennsylvania, Philadelphia,
Pennsylvania

Checked by: JOSEPH C. TOUCHSTONE, Department of Obstetrics and Gynecology,
Hospital of the University of Pennsylvania, Philadelphia, Pennsylvania

JEAN MARINO, Division of Biochemistry, Graduate Hospital of the
University of Pennsylvania, Philadelphia, Pennsylvania

Introduction

In 1950, Porter and Silber (1) described a highly sensitive quantitative reaction for cortisone (Δ^4-pregnene-17α, 21-diol-3, 11, 20-
trione) and related 17, 21-dihydroxy-20-ketosteroids. Nelson and
Samuels (2) subsequently devised a method for the purification of
extracts of human plasma, so that the levels of free corticoids might
be estimated by the application of the aforementioned reaction.
Both free and conjugated corticoids were found in human plasma
by means of techniques developed by Bongiovanni and Eberlein
(3).

In order to minimize interference by substances other than corticoids which are present in biological fluids, it is necessary to employ
some means of purification of extracts such as column chromatography, as described below. The "Porter-Silber" reagent consists
of phenylhydrazine in sulfuric acid. Presumably an osazone is
formed by reaction with the α-ketolic side chain. Since the phenylhydrazine reagent produces chromogens (with maximum absorption
at 410 mμ) only with steroids possessing 17, 21-dihydroxy-20-ketonic
functional groups, the presence of the 17-hydroxyl group is essential.
It has been suggested that a molecule of water is eliminated between carbons 16 and 17 in the course of this reaction and that the
osazone of Δ^{16}-steroids are responsible for the color produced. Thus,

* Based on the method of Nelson and Samuels (2).

it is apparent that compounds F and E, and their metabolites with a ketone at carbon 20, will react, whereas substances such as compound B will not. Since compound F and its metabolites predominate in human plasma, the recommended reagent is satisfactory for the measurement of adrenocortical function under most circumstances in man. Other reagents which measure the α-ketolic side chain in the absence of a hydroxyl group at carbon 17 may be employed.

The "Porter-Silber" reagent is extremely sensitive (E = 611,000 at 410 mμ for cortisone) and lends itself to the accurate estimation of less than 1 μg. of appropriate steroid, a quantity approximating that in 10 ml. of normal human plasma.

Reagents

1. Methylene chloride, redistilled. This solvent remains stable indefinitely. If chloroform is employed it must be repeatedly distilled immediately before use.

2. Absolute Ethanol (Gold Shield, Commercial Solvents Corp.). Redistillation is usually not required. However, if blank readings are high, redistillation from *m*-phenylenediamine HCl is recommended. One checker obtained good results with "Pharmco" alcohol (Publickers Industries, Inc.).

3. Phenylhydrazine HCl, recrystallized from ethanol 3 times.

4. Sulfuric acid (62.0% v/v).

5. Phenylhydrazine—sulfuric acid solution: (65 mg. of phenylhydrazine HCl in 100 ml. sulfuric acid). This is prepared immediately before use.

6. Florosil (60/100 mesh, Floridin Company, Tallahassee, Florida). Heat to 110° C. for 20–30 minutes shortly before use. Further treatment is not necessary.

NOTE: One of the greatest sources of error has been the loss of corticoids during chromatography. It is essential that the Florosil be dried by heating and that the column be well packed. Loose packing of the Florosil will lead to errors. It is advisable periodically to determine the recovery of pure compounds applied to columns; these should be constant and should yield 85% or more of the applied compound. Although the method of preparation of Florosil as described above is usually satisfactory, if nonspecific chromogens are formed with pure compounds or if the recovery of standards is too low, the original method for preparation of Florosil of Nelson and Samuels (2) should be used. The Florosil is washed with ethanol, *thor-*

oughly dried, and heated at 600°C. for 4 hours. Large batches may be prepared in this fashion, but it is still desirable to reheat the Florosil immediately before use.

7. Skellysolve B (Merck & Company).

8. β-Glucuronidase (Ketodase, Warner-Chilcott).

STANDARD SOLUTIONS

Cortisone or hydrocortisone in their free or acetylated forms are used as standards.

1. Stock standard. Use 50.0 mg. of cortisone or hydrocortisone per 100 ml. of absolute alcohol. Store in a brown, glass-stoppered bottle placed within a large jar containing ethanol-saturated cotton and keep at 4°–10° C.

2. Diluted standard. The 50.0 mg./100 ml. standard is diluted 1:100 and 1:25 with alcohol to give 5.0 and 20.0 μg./ml.; 0.25 ml. of these dilutions provide 1.25 and 5.0 μg. of standard per flask or, if related to a theoretical plasma sample of 10 ml., they are equivalent to 12.5 and 50.0 μg./100 ml.

Apparatus

1. Glass columns, 12 cm. long with an internal diameter of 10–11 mm. and a reservoir of approximately 75-ml. capacity, with or without stopcocks. These columns may be obtained from Kontes Glass Company, Vineland, New Jersey, Drawing No. 1063F.

2. Boiling flasks, round-bottom, approximately 100-ml. size, with small well blown in bottom, approximately 2.0-ml. capacity (Kontes, Drawing No. 1062F).

3. Microcuvettes, silica, 50 x 4.3 x 10 mm., approximately 1.2-ml. capacity (Pyrocell Manufacturing Company, New York City).

4. Beckman DU quartz spectrophotometer.

Obtaining and Storing Specimens

Draw blood (a volume slightly in excess of 20 ml.) into a syringe lubricated with about 0.5 ml. of liquid heparin (10 mg./ml.).

Centrifuge the sample within 15 minutes of collection and separate the plasma. The plasma sample should be treated at once as described below or stored at temperature below 0°C. Significant losses are encountered within 2–3 hours if the specimen is allowed to remain at higher temperatures.

Treatment of Specimens

1. Add 10.0 ml. of plasma slowly with agitation to 25.0 ml. of 95% ethanol. This operation is carried out at 45°C. and should take 3–4 minutes. After 15 minutes the precipitate is removed by centrifugation or filtration through Whatman No. 4 or No. 1 filter paper.

2. Extract the filtrate twice with 10 ml. of Skellysolve B in a separatory funnel. The Skellysolve B (upper layer) is discarded.

3. Evaporate the ethanolic solution almost to dryness with a current of air at a temperature not exceeding 45°C. in a 100-ml. flask.

4. Add 10 ml. of water and extract the aqueous solution three times with 10 ml. of methylene chloride. Pool the methylene chloride extracts from each specimen and evaporate as in step 3. *This residue contains the free corticoids and is treated as described under chromatography below.*

5. If it is desired to measure the conjugated corticoids, incubate the aqueous solution at 37°C. for 48 hours after adding 1.0 ml. of 1 M acetate buffer (pH 4.5) and 1.0 ml. of β-glucuronidase (5000 units).

6. Extract with 10 ml. methylene chloride three times as in step 4 and evaporate the pooled extracts as before. This residue contains the conjugated corticoids. The aqueous solution may now be discarded.

Column Chromatography

1. Insert a plug of glass wool with a glass rod into the lower ends of the glass columns supported by clamps on ring stands.

2. Add Florosil gradually by intermittent tamping with a glass or metal rod. The packing of the adsorbent in this fashion is essential. Add sufficient Florosil to a length of 70 mm. of tightly packed material. A plug of glass wool is gently introduced on the upper end of the column.

3. Pour 25–40 ml. of methylene chloride into the column and allow to drain slowly. When the fluid level is within 1 cm. of the top of the adsorbent, the residue prepared from the specimen extract previously dissolved in 5 ml. of methylene chloride is transferred to the column with a pipet. The column is at no time allowed to run dry.

4. When the fluid is within 1 cm. of the top of the adsorbent, pour into the column 25 ml. of the methylene chloride which was used to rinse the original flask containing the specimen residue. Add 25 ml. of 2% ethanol in methylene chloride. All solutions flowing from the column to this point are discarded.

5. When the previous fluid level is within 1 cm. of the top, pour onto the column 45 ml. of 25% ethanol in methylene chloride. Collect all the material flowing from the column at this time as a separate fraction. This fraction contains the corticoids purified for the colorimetric reaction.

6. Transfer the last fraction to a well-bottom flask and evaporate to dryness with a current of air at a temperature not above 45°C. Rinse down the flask once with 2–3 ml. of methylene chloride and again evaporate in order to concentrate all material in the well.

Colorimetric Procedure

1. To each residue obtained from the chromatography of plasma specimens in the well-bottomed flasks, add 0.25 ml. absolute ethanol and 0.75 ml. phenylhydrazine-sulphuric acid solution.

2. For the standards, pipet 0.25 ml. of each of the two diluted standards into small test tubes and add 0.75 ml. of the phenyl-hydrazine-sulphuric acid solution.

3. Prepare a blank containing 0.25 ml. of ethanol and 0.75 ml. of phenylhydrazine-sulphuric acid solution.

4. Place tubes and flasks in the dark at room temperature for 16 hours.

5. Transfer all specimens to the microcuvettes and read in the Beckman DU spectrophotometer at 370, 410, and 450 mμ, against the reagent blank.

Calculations

All readings are corrected in accordance with the principles of Allen (5) as folllows:

$$A \text{ (corrected at 410)} = A_{410} - \frac{A_{370} + A_{450}}{2}$$

The corrected readings for the two standards should be directly proportional to the concentrations within 2.0% and when plotted on a graph should provide a line which goes through zero.

A factor (F) is derived as follows:

$$F = \frac{\text{Corrected } A \text{ of standard}}{\text{micrograms of standard per tube}}$$

An average derived from the two standards is employed for this purpose.

The concentration of plasma corticoids is then calculated:

Micrograms of corticoids per 100 ml. plasma

$$= \frac{A \text{ (unknown corrected at 410 m}\mu)}{F} \times 10$$

Values for Healthy Subjects

Under basal conditions, between the hours 6 A.M. to 10 A.M., the free and conjugated corticoids are present in circulating plasma in mean concentrations of 8.0 ± 1.0 μg./100 ml. (range 2–18). Diurnal variations do occur so that reasonable standardization of the time of day at which specimens are obtained should be established. The value of this method as a measure of adrenocortical function is enhanced by repeating the estimation after the administration of adrenocorticotropin. Following the intravenous infusion of 0.25 units ACTH per kilogram body weight in saline over a 4-hour period, the failure to rise is indicative of adrenal insufficiency. In Cushing's syndrome (4) due to adrenal hyperplasia, the blood levels rise to unusually high levels (above 40.0 ug./100 ml.).

Sources of Error

Specimens which are inadequately purified may produce spurious colors which will not exhibit maximum absorption at 410 mμ. Under these conditions application of the correction factor will yield erroneously low values.

Certain drugs such as quinine and paraldehyde will influence the color reaction. It is advisable that all medication be suspended prior to obtaining specimens.

Modifications

If it is desired to measure only the free (unconjugated) plasma corticoids, a 10-ml. specimen may be directly extracted 3 times with 10 ml. of methylene chloride. The solvent is evaporated and the residue chromatographed on Florosil as above. Under these circumstances the Skellysolve-ethanol partition is not required. This procedure is practically identical to that of Nelson and Samuels (2).

Silber and Porter (6) have described a method wherein chromatography is not employed. There is some question as to the reliability of the chromogens produced with impure residues as an accurate measure of corticoids. If further improvements establish its validity, its simplicity should offer many advantages.

Methods employing paper chromatography for purification have also been described. Reagents other than phenylhydrazine have been employed for the colorimetric or fluorometric estimation of corticoids, some of which depend on somewhat different reactive groups on the steroid molecule. These have been reviewed elsewhere (7).

Smaller cuvettes may be employed so that only half the volume of reagents are used, as originally described by Nelson and Samuels (2). A Lowry-Bessey diaphragm attachment for the Beckman DU spectrophotometer is required for smaller cuvettes.

Peterson et al. (8) have recently devised a procedure for plasma corticoids which appears to be simpler in that no chromatography is required prior to color development. It has not been used for measuring conjugates.

17-Hydroxycorticosteroids in Urine

Since 99% of the 17-hydroxycorticosteroids found in urine are in the conjugated form, it is necessary to hydrolyze the conjugates and determine the total corticoids. The conjugated corticoids can be determined in the same manner as they are determined in plasma with the following modifications.

a. In step 6, the pooled methylene chloride extracts are washed with two 10-ml. portions of 0.1 N NaOH and two 10-ml. portions of water. These aqueous washings are discarded. The methylene chloride fractions are evaporated and chromatographed.

b. The residue from the chromatography is dissolved in 3.0 ml. of absolute ethanol. To a 1.0-ml. aliquot add 3.0 ml. of the phenylhydrazine-sulfuric acid solution. Prepare a blank containing 1.0 ml. ethanol and 3.0 ml. phenylhydrazine-sulfuric acid solution. Prepare standards containing 0.02 mg. and 0.06 mg. cortisone per 1.0 ml. alcohol (which are equivalent to 0.6 mg. and 1.8 mg. per 100 ml. of hypothetic urine processed as described above) and 3.0 ml. of the phenylhydrazine-sulfuric acid solution. These may be read in a micro- or regular cuvette.

c. If the chromatography is eliminated, the following simplified modification may be used. The dry residue from step 6 is dissolved in 3.0 ml. absolute ethanol. To one 1.0-ml. aliquot add 3.0 ml. phenylhydrazine-sulfuric acid solution; to another 1.0-ml. aliquot add 3.0 ml. 62.0% sulfuric acid solution. After the usual 16-hour color development, read each at 410 mμ against its own blank. The difference in absorbance between these two solutions is related to the concentration of hydroxycorticoids.

The answer obtained without chromatography may yield a higher result than the same sample chromatographed, since a small amount of chromogenic material may interfere. This artifact may be detected when a deep color is obtained in the portion of the sample treated with sulfuric acid. When this is the case, the more specific procedure using chromatography should be used.

The normal values for urine are 6–11 mg. per 24-hour period.

ADDENDUM. Further experience with two methods (6,8) since the preparation of this manuscript, indicates the advantage of simplicity and reliability

REFERENCES

1. Porter, C. C., and Silber, R. H., A quantitative color reaction for cortisone and related 17, 21-dihydroxy-20-ketosteroids. *J. Biol. Chem.* **185**, 201–207 (1950).
2. Nelson, D. H., and Samuels, L. T., A method for determination of 17-hydroxy-corticosteroids in blood. *J. Clin. Endocrinol. and Metabolism* **12**, 519–526 (1952).
3. Bongiovanni, A. M., and Eberlein, W. R., Determination, recovery, identification and renal clearance of conjugated adrenal corticoids in human peripheral blood. *Proc. Soc. Exptl. Biol. Med.* **89**, 281–285 (1955).
4. Grumbach, M. M., Bongiovanni, A. M., Eberlein, W. R., Van Wyk, J. J., and Wilkins, L., Cushing's syndrome with bilateral adrenal hyperplasia: A study of the plasma 17-hydroxycorticosteroids and the response to ACTH. *Bull. Johns Hopkins Hosp.* **96**, 116–125 (1955).
5. Allen, W. M., A simple method for analyzing complicated absorption curves, of use in the colorimetric determination of urinary steroids. *J. Clin. Endocrinol.* **10**, 71 (1950).
6. Silber, R. H., and Porter, C. C., The determination of 17, 21-dihydroxy-20-ketosteroids in urine and plasma. *J. Biol. Chem.* **210**, 923–932 (1954).
7. Bongiovanni, A. M., and Eberlein, W. R., Adrenocortical steroids in the peripheral blood of man. *J. Clin. Endocrinol. and Metabolism* **15**, 1524–1530 (1955).
8. Peterson, R. E., Karrer, A. and Guerra, S. L., Evaluation of Silber-Porter procedure for determination of plasma hydrocortisone. *Anal. Chem.* **29**, 144–149 (1957).

IRON IN SERUM*

Submitted by: Otto Schales, Department of Biochemistry, Tulane University School of Medicine and Alton Ochsner Medical Foundation, New Orleans, Louisiana

Checked by: Joseph V. Princiotto, Georgetown University School of Medicine, Washington, D. C.

    Robert S. Melville, Department of Biochemistry, Veterans Administration Hospital, Iowa City, Iowa

Introduction and Principles

In 1925, Barkan (1–3) reported that serum contains a small amount of iron. Later, Barkan and Schales (4) found that this iron is bound to the globulin fraction of the serum proteins, from which it is readily split off on mild acidification. The iron-binding compound is found in subfraction IV-7, which consists chiefly of a β_1-globulin with a relatively low molecular weight near 90,000 (5). Barkan described the probable physiological role of this protein-bound iron for the transport of iron in the body, and later work essentially confirmed his views.

In his earlier studies, Barkan determined iron by measuring the color intensity of ether solutions of ferrithiocyanate, obtained by adding potassium thiocyanate to protein-free serum filtrates and extracting with peroxide-containing ether in glass-stoppered tubes. (Peroxide was necessary to convert all iron present to the trivalent form.) This early method is today of historical interest only and has been replaced by simpler, more sensitive, and more convenient procedures. In the method to be described here, which was first studied by Heilmeyer and Plötner (6), 1,10-phenanthroline is used as reagent and the light absorption of its red-colored complex with ferrous iron is measured at its absorption maximum (510 mμ). The color of the phenanthroline-iron complex is stable and the relationship between absorbance and concentration is linear.

* Based on the methods of Heilmeyer and Plötner (6), Schales (7), and Barkan and Walker (8).

69

Heilmeyer's measurements were carried out with the accurate but expensive visual Koenigs-Martens spectrophotometer. When simple photoelectric colorimeters became available commercially, modifications of Heilmeyer's method were published by Schales (7) and by Barkan and Walker (8). Since then, many articles, too numerous to mention here, have dealt with the use of 1,10-phenanthroline as reagent for the determination of small amounts of iron in a variety of media. All of these methods have in common decreased acidity of the iron-containing solution, so that the pH falls between 3 and 9. If the total iron, and not just the ferrous iron, is to be determined, a reducing agent is added before producing color with the complexing reagent. The method described here has proved satisfactory for many years in the submitter's laboratory and differs in minor details only from the procedures described in the references.

Special Handling of Glassware

It is essential, of course, that all glassware used in the determination of iron be cleaned carefully by immersion in bichromate-sulfuric acid solution overnight, followed by washing with hot water and generous rinsing with distilled water. Such glassware is best set aside for iron determinations only and must be protected carefully from contamination after it has been cleaned and dried.

Reagents

1. HCl (approximately 0.35 N). Make by diluting 30 ml. concentrated HCl to 1 l. with water.

NOTE: Titration of a sample used in our tests showed that the acid was 0.347 N. This acid is of sufficient strength to split off the iron from its combination with globulin. Various investigators have used HCl of different normality for the same purpose, even as high as 6 N. It should be kept in mind that higher acid concentrations than that recommended here require more neutralizing agent to reduce the pH to the range for optimal color.

2. Trichloroacetic acid solution, 20 g./100 ml.

NOTE: Instead of removing protein by precipitation with this reagent, some authors have used wet ashing. This is not recommended by the submitter, since serum always contains traces of hemoglobin, no matter how carefully the specimen was collected. Ashing liberates the iron from hemoglobin and this results in erroneously high serum iron values. The presence of 0.1% hemolysis (16 mg. hemoglobin contains 53.6 μg. iron) would increase "serum iron" by about 50% over the true value. Heilmeyer and Plötner (6) have shown that the treatment with HCl and trichloroacetic acid does not liberate iron from hemoglobin, even if there is considerable hemolysis. Barkan's data prove that no iron is lost with the trichloroacetic acid precipitate, since the results obtained were identical with those found in ultrafiltrates of HCl-treated serum. Furthermore, it has been shown by Heilmeyer and Plötner (6) that the iron content after ashing minus the hemoglobin iron (calculated from an independent hemoglobin determination) equals the iron content of the trichloroacetic acid filtrate.

3. Potassium acetate solution, 50 g./100 ml.

NOTE: This reagent, instead of the sodium or ammonium salt is recommended, since it is more soluble than crystalline sodium acetate and far less hygroscopic than the ammonium salt.

4. Hydroquinone solution, 1 g./100 ml.

NOTE: Other reducing agents which may be substituted are hydroxylamine hydrochloride, hydrazine, and ascorbic acid. Hydroquinone has an advantage in that the appearance of a brownish color indicates that its useful life is over and that a fresh solution should be prepared.

5. 1,10-phenanthroline solution (o-phenanthroline hydrate), 0.1 g./100 ml.

STANDARD IRON SOLUTIONS

Stock Solution

This contains 1.00 mg. iron per 100 ml. Make by dissolving 70.2 mg. ferrous ammonium sulfate, $Fe(NH_4)_2$ $(SO_4)_2 \cdot 6H_2O$, in water containing 0.2 ml. 2 N H_2SO_4 and making to a final volume of 1000 ml.

Working Standard

Make by diluting 20 ml. stock solution to 100 ml. with water. This working standard contains 200 μg. iron per 100 ml. A series of working standards for calibrating photoelectric colorimeters may be prepared by suitable dilutions of the stock solution.

Procedure

To 2.00 ml. of serum, or 2.00 ml. of water, or 2.00 ml. of working standard solution, add 1.0 ml. of 0.35 N HCl and let stand for 1 hour. Then add 1.0 ml. of 20% trichloroacetic acid and let stand for 15 minutes. At the end of this period, centrifuge (the serum sample only) for about 15 minutes at about 2000 r.p.m.

NOTE: A blank is absolutely essential. Though all reagents used are colorless, they always contain traces of iron. Tests in the submitter's laboratory, using reagents of highest purity, gave iron values in blanks, which, if neglected, would have raised the serum iron result by 12 μg./100 ml.

Transfer 2.00 ml. of the clear supernatant (or mixture) to a reaction vessel, add 0.25 ml. of a 50% potassium acetate solution followed by 0.15 ml. of a 1% hydroquinone solution and finally by 0.5 ml. of a 0.1% 1,10-phenanthroline solution. Let stand until the pink, stable color is fully developed (about 20 minutes) and read in a 10.0-mm. cuvette in a Beckman spectrophotometer at the wavelength of maximum absorption, i.e. at 510 mμ.

NOTE: The addition of potassium acetate resulted, in the submitter's laboratory, in a pH of 4.5 for the blank and 4.7 for the serum sample. This difference has no influence on the final color intensity, which remains constant from pH 3 to 9.

Be sure to determine the wavelength of maximum absorption on your instrument. Errors in the calibration of spectrophotometers coming from the factory of as much as 8 mμ have been observed by the submitter. Taking the readings 2 mμ on either side of the maximum is permissible, since it will cause an error of less than 1 μg. iron per 100 ml. serum. However, taking a reading 10 mμ to the longer wavelength side (520 mμ) or 20 mμ to the shorter wavelength side (490 mμ) will result in an error of 10% if the calibration factor determined at the wavelength of maximum absorption is used.

Calibration

Heilmeyer and Plötner (6), using a visual spectrophotometer, determined the absorbance index a_s (1 g. iron per 100 ml., 10.0-mm. light path) as 2054. From this value one can calculate a molar absorbance index a_M (55.85 g. iron per 1000 ml., 10.0-mm. light path) of 11,470. Later investigators, using photoelectric spectrophotometers, found slightly lower constants, such as 11,100 (9), 11,080 (10), and 11,000 (11). Values obtained in the submitter's laboratory, with standard solutions ranging from 100 to 300 μg. iron per 100 ml., fell within the limits from 11,160 to 11,510 and gave an average index of 11,320, which is in reasonable agreement (1.3%) with the value found by Heilmeyer and Plötner.

From a molar absorbance index of 11,320, the absorbance index a_s (1 g. iron per 100 ml.) is calculated as 2027. Each investigator should re-establish the value by submitting a number of standard iron solutions to the test procedure. The working standard solution mentioned under reagents, which contains 200 μg. iron per 100 ml. and which is diluted 2.9-fold by the addition of the reagents, should, for example, have a net absorbance A_s (corrected for the blank) of $2027/(5 \times 2.9 \times 1000) = 0.140$. In the submitter's laboratory, this value ranged from 0.1380 to 0.1421, averaging 0.140.

Calculation

$[A_s$ (unknown) $- A_s$ (blank)] x 1430 = Micrograms Fe/100 ml. serum.

The calibration factor $F = 1430$ is based on a molar absorbance index of 11,320 and reflects the fact that the original serum was diluted 2.9-fold by the addition of reagents. It is calculated with the equation:

$$F = \frac{5585 \times 1000 \times 2.9}{11,320} = 493.4 \times 2.9 = 1430$$

If a dilution different from 2.9-fold is used, the corresponding calibration factor is obtained by substituting the new dilution factor for the value 2.9 used above.

Calibration of Photoelectric Colorimeters

The use of a precision spectrophotometer has the advantage that a known absorbance index is available for the calculation and that only a small sample is required (2.9 ml. final volume), which may be further reduced by the use of microcuvettes with 10.0-mm. light path (to less than 1 ml.).

However, frequently only photoelectric colorimeters are at the disposal of the analyst and must serve instead of a spectrophotometer. To test the suitability of a typical photoelectric colorimeter, the submitter calibrated a Lumetron Colorimeter, Model 402-E, using filter M-515. Test tubes with a diameter of 18.5 mm. were used as cuvettes. These tubes were matched, so that a solution, giving in the master tube a transmittance of 50.0%, would, in any of the other tubes, give readings deviating from the master-tube reading by not more than 0.2% transmittance (49.8–50.2%). This corresponds to maximal absorbance differences (when reading the same solution in different tubes of the set) of ±0.6%.

Since such tubes require 8 ml. of solution for reading, three times as much starting material as was needed for the spectrophotometric procedure had to be employed. The corrected absorbances were plotted against iron concentrations (the submitter used concentrations ranging from 100 to 300 μg./100 ml.) yielding a straight line, going through the origin, with an increase in absorbance (slope) per 50μg. iron per 100 ml. of 0.042 and resulting in a calibration factor of 1190. Individual values varied from this average by not more than ±1.5%. The value of the calibration factor changes, of course, from one instrument to the next (even of the same model) and must be determined carefully by each investigator, using a reasonable number of different standard solutions in duplicate (the submitter used 7).

Discussion

Serum Iron in Health and Disease

Numerous investigators during the last thirty years, using a variety of methods, have generally agreed that the normal range of serum iron is approximately (micrograms per 100 ml. serum) in men, 80–160 (average 125); and in women, 65–130 (average 90). Sometimes, "normal values" considerably in excess of these figures have been reported, but these higher values were in error and were due either to contaminated glassware and reagents or to an ashing procedure which included hemoglobin iron in the result.

Low values for serum iron are encountered in hemorrhagic and hypochromic types of anemia. Moore and his associates (12) found in 11 patients with hypochromic microcytic anemia values ranging from 14 to 46 μg./100 ml., as contrasted to their normal values of 122 μg. in males and 98 μg. in females. Powell (13) reported values from 25 μg. to 90 μg./100 ml. serum (average 47 μg.) in patients suffering from the same condition; her average normal values were higher than those of other investigators, namely 143 μg./100 ml. in males and 117 μg./100 ml. in females. Low serum iron levels were also found in patients with various infections (6), even in people afflicted with a mild common cold, as observed by Venndt (14), who studied this problem under the submitter's guidance.

High values for serum iron (150–350 μg./100 ml.) were found in forms of anemia characterized by diminished hemoglobin formation not due to iron deficiency, as in untreated pernicious anemia (12). Liver therapy, leading to increased utilization of iron for hemoglobin synthesis, resulted in a fall of the serum-iron concentration in these patients below the normal range (6, 12). High serum-iron values were also seen in thalassemia (Cooley's anemia or erythroblastic anemia). Smith and co-workers (15) reported an average of 270 μg./100 ml. in a series of 15 children with this disease. Reports on serum iron in hemochromatosis are somewhat conflicting, but recent data indicate that significant abnormalities may occur not infrequently (16, 17) and elevated serum-iron concentrations ranging from 144–282 μg./100 ml. have been found. Rath and Finch (16) found an average of 224 μg./100 ml. in 9 cases of hemochromatosis, as compared with their average normal value of 100 μg./100 ml. Of importance as an aid in the diagnosis of hemochromatosis is the iron-binding capacity of serum or plasma.

Iron-binding Capacity of Serum

Holmberg and Laurell (18) noticed that ferrous iron, when added to serum, became firmly bound. If the amount added was just sufficient to saturate the serum, or was less, no reaction for ferrous ions was obtained with bipyridyl. However, the addition of Fe^{++} above that limit resulted in a reaction with bipyridyl and permitted the photometric determination of the unbound iron. The authors found that normal serum can bind a total of about 300 μg. of iron per 100 ml., which means that normal serum, as obtained from a patient, containing about 100 μg./100 ml. can bind an additional 200 μg./100 ml. and is only about one-third saturated in respect to iron. Holmberg and Laurell also reported that untreated patients with pernicious anemia often had such high serum-iron values that the capacity to bind additional iron was absent.

Schade and Caroline (19), who characterized the iron-binding globulin in plasma, observed that the colorless protein turned salmon-red when combined with iron (absorption broad, from 420–530 mμ, maximum at 465 mμ) and used this property for the colorimetric determination of the iron-binding capacity of plasma. This method was also employed by later investigators (16, 20). A semiquantitative procedure may be described as follows (16): small increments (0.05 ml.) of an iron standard solution (14 mg. ferrous ammonium sulfate, plus 0.5 ml. N acetic acid in 100 ml. total volume; 1 ml. = 20 μg. iron) are added to 2 ml. serum, which has been diluted with 3 ml. 0.9% NaCl solution. After each addition, light transmittance at 525 mμ is measured, using a Beckman spectrophotometer. As long as the addition of iron produces more of the pink globulin-iron complex, transmittance decreases. When the protein is saturated, further addition of iron does not result in a further decrease in transmission. This method is rather crude, since each increment of 0.05 ml. corresponds to an increase of 50 μg. iron per 100 ml. serum. The wavelength of 525 mμ was used to avoid interference by other components in serum. However, Cartwright and Wintrobe (20) selected 490 mμ as suitable for these measurements.

Simple as these methods are, they confirmed the original data of Holmberg and Laurell that the total iron-binding capacity of serum is about 300 μg/100 ml., of which about 100 μg./100 ml. is present as normal serum iron. A more accurate method, using radioactive iron for determining the iron-binding capacity of serum, has been described recently by Feinstein et al. (21). These authors add radioactive iron to serum, let stand for 15 minutes, and precipitate the protein-iron complex with neutral, saturated ammonium sulfate solution (4). The radioactive iron in the filtrate is measured and subtracted from that originally added; this gives the value for the bound radioactive iron. The iron-binding capacity can then be calculated with the help of the ratio of total iron to radioactive iron in the standard solution. Saturated ammonium sulfate solution contains sufficient iron to prevent the use of ordinary iron in this method. The authors have established that the protein-bound radioactive iron is not displaced by the stable iron introduced with the ammonium sulfate.

In hemochromatosis, the iron-binding capacity averages only about 25 μg. iron per 100 ml. (additional uptake of iron, not including the amount of iron already present), as contrasted to a binding capacity of 200 μg./100 ml. normal plasma. This means that the capacity of the serum to bind iron is apparently saturated to the extent of about 90% in this disease, as contrasted to the normal 30–35% saturation. This abnormal situation is further illustrated by the fact that administration of iron results in a much smaller rise in serum iron in patients with hemochromatosis than in normal individuals.

A Substituted Phenanthroline as Reagent for Serum-Iron Determinations (Bathophenanthroline)

Recently, 4,7-diphenyl-1,10-phenanthroline has been suggested as a reagent for iron determinations (9, 22), since its iron complex is twice as intensely colored (molar absorbance index = 22,400 at 533 mμ) than that with 1,10-phenanthroline. In view of the small amounts of iron found in serum, a reagent with greater sensitivity is highly desirable. However, the submitter feels that more work needs to be done with this new reagent before it can take the place of 1,10-phenanthroline. In particular, it will be necessary to clarify the reason for the startling normal values obtained with this compound (22), which are considerably higher and in sharp contrast to the rather uniform results reported by numerous investigators with a variety of other methods.

REFERENCES

1. Barkan, G., Eisenstudien, 1. Mitteilung. Zur Frage der Einwirkung von Verdauungsfermenten auf das Hämoglobineisen. Z. physiol. Chem. **148,** 124–154 (1925).
2. Barkan, G., Eisenstudien, 3. Mitteilung. Die Verteilung des leicht abspaltbaren Eisens zwischen Blutkörperchen und Plasma und sein Verhalten unter experimentellen Bedingungen. Z. physiol. Chem. **171,** 194–221 (1927).
3. Barkan, G., Eisenstudien, 6. Mitteilung. Über Bestimmungsmethodik und Eigenschaften des "leicht abspaltbaren" Bluteisens. Z. physiol. Chem. **216,** 1–16 (1933).
4. Barkan, G., and Schales, O., Chemischer Aufbau und physiologische Bedeutung des "leicht abspaltbaren" Bluteisens. Z. physiol. Chem. **248,** 98–116 (1937).
5. Edsall, John T., The plasma proteins and their fractionation. Advances in Protein Chem. **3,** 383–479 (1947).
6. Heilmeyer, L., and Plötner, K., "Das Serumeisen und die Eisenmangelkrankheit." Gustav Fischer, Jena, 1937.
7. Schales, O., The determination of serum iron. Klett-Summerson Photoelectric Colorimeter Manual, Klett Mfg. Co., New York, 1939.
8. Barkan, G., and Walker, B. S., Determination of serum iron and pseudohemoglobin iron with o-phenanthroline. J. Biol. Chem. **135,** 37–42 (1940).
9. Moss, M. L., Mellon, M. G., and Smith, G. F., Color reactions of 1,10-phenanthroline derivatives. Anal. Chem. **14,** 931–933 (1942).
10. Mehlig, J. P., and Hulett, H. R., Spectrophotometric determination of iron. Anal. Chem. **14,** 869–871 (1942).
11. Peterson, R. E., Improved spectrophotometric procedure for the determination of serum iron. Anal. Chem. **25,** 1337–1339 (1953).

12. Moore, C. V., Doan, C. A., and Arrowsmith, W. R., Studies in iron transportation and metabolism. II. The mechanism of iron transportation: its significance in iron utilization in anemic states of varied etiology. *J. Clin. Invest.* **16,** 627–648 (1937).
13. Powell, J. F., Serum iron in health and disease. *Quart. J. Med.* **13,** 19–26 (1944).
14. Venndt, H., personal communication (1938).
15. Smith, C. H., Sisson, T. R. C., Floyd, W. H., Jr., and Siegal, S., Serum iron and iron binding capacity of the serum in children with severe Mediterranean (Cooley's) anemia. *Pediatrics* **5,** 799–807 (1950).
16. Rath, C. E., and Finch, C. A., Chemical, clinical, and immunological studies on the products of human plasma fractionation. 38. Serum iron transport, measurement of iron-binding capacity of serum in man. *J. Clin. Invest.* **28,** 79–85 (1949).
17. Houston, J. C., and Thompson, R. H. S., The diagnostic value of serum iron studies in hemochromatosis. *Quart. J. Med.* **21,** 215–224 (1952).
18. Holmberg, C. G., and Laurell, C. B., Studies on the capacity of serum to bind iron. *Acta Physiol. Scand.* **10,** 307–319 (1945).
19. Schade, A. L., and Caroline, L., An iron binding component in human blood plasma. *Science* **104,** 340–341 (1946).
20. Cartwright G. E., and Wintrobe, M. M., Chemical, clinical, and immunological studies on the products of human plasma fractionation. 39. The anemia of infection. Studies on the iron-binding capacity of serum. *J. Clin. Invest.* **28,** 86–98 (1949).
21. Feinstein, A. R., Bethard, W. F., and McCarthy, J. D., A new method, using radioiron, for determining the iron-binding capacity of human serum. *J. Lab. Clin. Med.* **42,** 907–914 (1953).
22. Kingsley, G. R., and Getchell, G., Serum iron determination. *Clin. Chem.* **2,** 175–183 (1956).

17-KETOSTEROIDS IN URINE*

Submitted by: ALBERT L. CHANEY, Albert L. Chaney Chemical Laboratory, Glendale, California.

Checked by: NELSON F. YOUNG, Medical College of Virginia, Richmond, Virginia
ROBERT L. DRYER, University Hospitals, State University of Iowa, Iowa City, Iowa

Introduction

The group of compounds known as 17-ketosteroids are metabolites of hormones produced by the adrenal cortex, the testes, and to a limited extent, the ovaries. Their determination is of clinical value in evaluating disturbances of function of these endocrine glands.

The steroids which have a keto group at carbon atom number 17 react with *m*-dinitrobenzene and alkali in alcoholic solution to produce a purple color. This reaction was first applied to steroid determinations by Zimmermann (1) and subsequently developed by Callow *et al.* (2) and Holtorff and Koch (3). Many variations in technique have been used by subsequent authors, and recently micromodifications in which small aliquots of urine are used have been described by Vestergaard (4), Drekter *et al.* (5), and Klendshoj *et al.* (6).

The method of Drekter *et al.* (5) has been selected as one which combines relative simplicity of technique with good reproducibility, and in which interfering pigments are reduced to a minimum.

Principles

The ketosteroids are excreted in the conjugated form as glucosiduronates and sulfates. Hydrolysis by strong acid at elevated temperatures is used to produce the free steroids. In the free state they are readily extracted by a variety of organic solvents, of which ethylene dichloride has several advantages. Interfering pigments and other compounds reacting with the Zimmermann reagents to

---

* Based on the method of Drekter, Heisler, Scism, Stern, Pearson, and McGavack (5).

79

produce colored products may be removed by shaking the ethylene dichloride extract with pellets of solid sodium hydroxide. An aliquot of the purified extract is evaporated, and after color development with *m*-dinitrobenzene and alkali under controlled conditions of temperature and time, the purple color is measured photometrically.

Reagents

1. Hydrochloric acid, concentrated, reagent grade A. C. S.

2. Ethylene dichloride, technical grade, redistilled.

3. Sodium hydroxide, pellets, reagent grade A. C. S. Protect from moisture and carbon dioxide absorption.

4. m-Dinitrobenzene. Purify by recrystallization and freshly dissolve as a 1.0% solution in ethyl alcohol.

NOTE: Purification of *m*-dinitrobenzene [method of Callow *et al.* (2)]: Dissolve 20 g. of *m*-dinitrobenzene in 750 ml. of 95% ethanol at 40°C. Add 100 ml. of 2 *N* sodium hydroxide, allow to stand for 5 minutes, and cool. Then precipitate the *m*-dinitrobenzene by diluting with 2500 ml. of water. Filter the precipitate on a Buchner funnel, wash thoroughly with water, and recrystallize from 100 ml. of hot 95% ethanol.

A test of satisfactory purity may be made by setting up a blank with the same proportions of *m*-dinitrobenzene and sodium hydroxide solutions as in the standard test. After color development and addition of 75% ethanol, the optical density should not exceed 0.05. The increase in the optical density of the blank may be due either to the quality of the alcohol or to the deterioration of the *m*-dinitrobenzene. The 1% solution of *m*-dinitrobenzene may be kept for a few days in the dark and refrigerated, but the blank should be tested daily.

5. Potassium hydroxide, reagent grade A. C. S. Prepare as 8.0 ±0.2 *N* aqueous solution and check by titration (approximately 54 g. of potassium hydroxide per 100 ml. of solution).

6. Alcohol, ethyl, 95%, or absolute, and 75% (free from ketones or other impurities causing color with Zimmermann reagents). Dilute the 95% alcohol with water to prepare the 75% concentration.

NOTE: Purification of alcohol [p. 501 of reference (11)]: Add approximately 4 g. of *m*-phenylenediamine hydrochloride to a liter of ethyl alcohol. Allow to stand in the dark for a week with occasional shaking. Redistil in an all-glass apparatus, discarding the first and last portions.

7. Dehydroepiandrosterone, 60 mg./liter of alcohol. Use as the steroid standard.

Apparatus and Equipment

Glass-stoppered centrifuge tubes: 50–60-ml. capacity.

Test tubes: 18 mm. x 100 mm.

Pipets for measuring 20 ml., 3 ml., 0.4 ml., and 0.3 ml., and of 0.5% or less error.

Funnels: 2½-inch diameter.

Filter paper: 11-cm. diameter, Whatman No. 1 or No. 43.

Shaking machine: Kahn shaker or equivalent, adapted to shaking 50-ml. centrifuge tubes.

Water bath or equivalent constant-temperature container for 25°C.

Photometer: filter photometer and 525 mμ. filter and cuvette requiring not more than 3.5 ml. of solution, or spectrophotometer which can be set at 520 mμ.

Sample

Urine should be collected over a complete 24-hour period, as from 7:00 A.M. to 7:00 A.M. The specimen should be kept cool during collection and is preferably preserved by collection in a container to which has been added sufficient hydrochloric acid to give a final concentration of about 1%.

Procedure

Measure the 24-hour volume of urine in milliliters and record. Place 20.0 ml. of the sample in a 50-ml. glass- or polyethylene-stoppered centrifuge tube. Add 4 ml. of concentrated hydrochloric acid. Place the test tube in a water bath at 100°C. for 10 minutes. Cool promptly to room temperature. Add an equal volume (20.0 ml.) of ethylene dichloride and place the test tube in the shaking machine for 10 to 15 minutes. If necessary centrifuge the solution briefly to obtain a separation of layers. Remove the aqueous layer by aspiration and pour a 12–15-ml. aliquot of the ethylene dichloride layer into a dry Whatman No. 43 filter and collect in a dry 50-ml. glass-stoppered centrifuge tube. In an additional centrifuge tube, place an identical amount of ethylene dichloride to serve as the reagent blank. To each tube add 3 g. of sodium hydroxide pellets and place the tubes in the shaking machine for 10 minutes. Using the Whatman No. 43 filter paper, filter the solutions to remove the pellets and insoluble reaction products. The funnel should be covered with tin foil or a watch glass to prevent evaporation and change in concentration. Transfer 3.00-ml. aliquots of the ethylene dichloride filtrates (samples and blank) to appropriately labeled test tubes. This is equivalent to 3-ml. aliquots of the original specimen. Evaporate the filtrates to dryness in a water bath. Cool the tubes and add 0.1 ml. of absolute alcohol to each tube. Evaporate to dryness in the water bath, letting the alcohol reflux and wash the residue to the bottom of the tube. Remove the last traces of solvent vapor with a slow stream of air. To each tube add 0.40 ml. of the 1% m-dinitrobenzene solution and 0.30 ml. of 8.0 N potassium hydroxide. Mix thoroughly and allow the tubes to stand for 25 minutes at 25°C. in a water bath. Add 3.0 ml. of 75% alcohol to each tube and mix thoroughly. Measure the absorbance of the blank and sample with a filter photometer or spectrophotometer. The optimum wavelength is 520 mμ.

Standardization

Dehydroepiandrosterone is readily available in suitable purity and is a reproducible standard.

Prepare a stock standard containing 60.0 mg./liter of dehydroepiandrosterone in absolute alcohol. In 3 test tubes evaporate 0.50, 1.00, and 1.50 ml. of standard solution in a water bath. Add 0.40 ml. 1% m-dinitrobenzene and 0.30 ml. 8.0 N KOH, and develop color as described for sample. These standards are equivalent to 10, 20, and 30 mg. of steroid per liter of hypothetical urine when the method is performed as described.

Calculations

The following calculation applies when 1.00 ml. of standard is used containing 0.060 mg. dehydroepiandrosterone and 3 ml. of extract corresponding to 3 ml. of urine sample.

Concentration of 17-ketosteroids, (mg./100 ml.)

$$= \frac{\text{Absorbance of specimen} - \text{absorbance of blank}}{\text{Absorbance of standard} - \text{absorbance of blank}}$$

$$\times 0.060 \text{ mg.} \times \frac{100 \text{ ml.}}{3 \text{ ml.}}$$

$$17\text{–Ketosteroids, (mg./24 hr.)} = \frac{\text{Conc. mg./100 ml.}}{100 \text{ ml.}} \times 24\text{-hr. vol. (ml.)}$$

(Calculated as dehydroepiandrosterone)

Discussion

The 17-ketosteroids found in urine comprise a large number of compounds and there are at least six which are normally present in quantities of 1 mg. or more in a 24-hour specimen (7, 8). They are present originally as water-soluble conjugates (sulfates and glucosiduronates), which can be hydrolyzed to free steroids by various means. Since the various conjugates differ markedly in the ease with which they may be split, and also since some of the steroids are destroyed or altered by the hydrolysis, no single method of hydrolysis has been found which gives both complete hydrolysis and 100% recovery of the original steroids.

The determination of this heterogeneous group of substances, which are present in varying proportions in different individuals accordingly requires a series of compromises, in which reproducibility of results, rather than complete quantitative recovery, is a primary consideration.

These compromises involve first, the conditions of hydrolysis, secondly, the removal of interfering substances which give colors with the Zimmermann reagents, and lastly, the conditions for color development.

A very large variety of hydrolytic procedures has been proposed, each of which has its advantages for particular steroid conjugates. Hydrolysis at elevated temperatures with strong acid is the most generally useful, its principal disadvantage being its destructive action on β-ketosteroids, such as dehydroepiandrosterone.

Enzyme hydrolysis by glucuronidase liberates only 30–60% of the ketosteroids, since it affects glucosiduronates only. This hydrolysis requires 24–48 hours.

The conditions selected here for hydrolysis are typical of the procedures which depend on strong acid at boiling temperature for short periods of time.

The present information about hydrolysis is well summarized by Lieberman et al. (9).

A large number of solvents are available following hydrolysis for extraction of the steroids. Desirable characteristics are favorable distribution coefficient for steroids, relative nontoxicity, ease of subsequent evaporation, and low solubility for nonsteroid materials, interfering pigments, etc. These are well met by ethylene dichloride.

The originators of the procedure (5) have shown that shaking with solid sodium hydroxide pellets removes to a great extent the nonsteroid pigment-forming substances extracted by ethylene dichloride.

The Zimmermann color reaction (1–3) given by compounds containing the methylene ketone grouping—CH_2CO— is obtained by treatment of the residue with alcoholic solution of m-dinitrobenzene in the presence of alkali. The intensity of color produced is dependent on several factors: the relative proportions of alcohol and water in the reaction mixture, the strength of alkali, the temperature and the length of time of color development. All of these factors therefore require careful control and standardization.

When interfering pigment-forming materials are at a minimum, as in this procedure, the choice of these conditions is largely a matter of relative convenience, once they are standardized. The rate of color formation is slow and not complete in the 25 minutes' time interval specified, so timing should be carefully observed. In general, rate of color formation is accelerated by increasing alkali strength or by rise of temperature. Some workers prefer alcoholic solutions of alkali because of greater specificity of color, but when interfering pigments are at a minimum this advantage disappears and lack of stability is a serious disadvantage.

Additional review material on methods of determination may be found in recent publications (8–11).

Reproducibility of the Results

When duplicate determinations are made on the same extract, the reproducibility of the Zimmermann color reaction should be within 2 or 3% in the normal range of optical densities. The hydrolysis stage and preparation of the extract should not materially increase the variability. Inherent errors in the determination result from the physiological variation of excretion from day to day and also at different times of the day. Another source of error is the inaccurate collection of the 24-hour specimen. When it contains acid and is stored in the refrigerator, the original specimen is reasonably stable for periods up to 30 days [p. 483 of reference (11)].

Interpretation of Results

Healthy adult males excrete 10–30 mg. per 24 hours and females, 5–14 mg. In patients having diseases of the adrenal glands, these values may be increased markedly or may be diminished almost to zero, depending upon the disease state. In normal children under 5, the excretion is less than 2 mg. per 24 hours and gradually increases during later childhood and adolescence to adult values. Normal values utilizing this procedure have been published by Kenigsberg *et al.* (12). The interpretation of abnormal values is discussed in considerable detail in a recently published monograph (13).

References

1. Zimmermann, W., A color reaction of sex hormones and its application to colorimetric determination. *Z. physiol. Chem.* **233**, 257–264 (1935); Colorimetric determination of sexual hormones. *Z. physiol. Chem.* **245**, 47–57 (1936).

2. Callow, N. H., Callow, R. K., and Emmens, C. W., Colorimetric determination of substances containing the grouping—CH_2CO—in urine extracts as an indication of androgen content. *Biochem. J.* **32**, 1312–1331 (1938).

3. Holtorff, A. F., and Koch, F. C., The colorimetric estimation of 17-ketosteroids and their application to urine extracts. *J. Biol. Chem.* **135**, 377–392 (1940).

4. Vestergaard, P., Rapid micro-modification of Zimmermann-Callow procedure for determination of urinary 17-ketosteroids. *Acta Endocrinol.* **8**, 193–214 (1951).

5. Drekter, I. J., Heisler, A., Scism, G. R., Stern, S., Pearson, S., and McGavack, T. H., The determination of urinary steroids. I. The preparation of pigment-free extracts and a simplified procedure for the estimation of total 17-ketosteroids. *J. Clin. Endocrinol. and Metabolism* **12**, 55–65 (1952).

6. Klendshoj, N. C., Feldstein, M., and Sprague, A., Determination of 17-ketosteroids in urine. *J. Clin. Endocrinol. Metabolism* **13**, 922–927 (1953).

7. Lieberman, S., and Dobriner, K., Steroid excretion in health and disease: I. Chemical aspects. *Recent Progr. in Hormone Research* **3**, 71–101 (1948).

8. Mason, H. L., and Engstrom, W. W., The 17-ketosteroids: Their origin, determination, and significance. *Physiol. Revs.* **30**, 321–374 (1950).

9. Lieberman, S., Mond, B., and Smyles, E., Hydrolysis of urinary ketosteroid conjugates. *Recent Progr. in Hormone Research* **9**, 113–134 (1954).

10. Munson, P. L., and Kenny, A. D., Colorimetric analytical methods for neutral 17-ketosteroids of urine. *Recent Progr. in Hormone Research* **9**, 135–161 (1954).

11. Engel, L. L., The assay of urinary neutral 17-ketosteroids. *In* "Methods of Biochemical Analysis" (D. Glick, ed.), Vol. 1, pp. 479–509. Interscience, New York, 1954.

12. Kenigsberg, S., Pearson, S., and McGavack, T. H., The excretion of 17-ketosteroids. I. Normal values in relation to age and sex. *J. Clin. Endocrinol.*, **9**, 426–429 (1949).

13. Dorfman, R. I., and Shipley, R. A., Androgens: Biochemistry, Physiology, and Clinical Significance. Wiley, New York, 1956.

PANCREATITIS-LIPASE[*][†]

Submitted by: RICHARD J. HENRY, Bio-Science Laboratories, Los Angeles, California

Checked by: HARWOOD M. TAYLOR, Department of Biochemistry, Duke University School of Medicine, Durham, North Carolina

Introduction and Principle

This method supersedes the method for "lipase" employing tributyrin as substrate which appeared in Volume I. It has now been established (1, 2) that "tributyrinase" is *not* elevated in acute pancreatitis, the disease for which a "lipase" determination is used as a diagnostic aid.

The use of olive oil as a substrate for measuring lipolytic activity in pancreatic extracts can be traced back into the nineteenth century. In 1911, Rona and Michaelis (3) reported the use of tributyrin as a substrate and the determination of the degree of hydrolysis by stalagmometric measurements. Cherry and Crandall (4), in 1932, described the application of the olive oil procedure to serum; they found that, following pancreatic duct ligation in dogs, "lipase" increased if olive oil was used as substrate but not when tributyrin was used. That there is increased lipase activity in serums from patients with pancreatitis using olive oil as substrate has been repeatedly confirmed (1, 2, 5–8).

In 1948, Goldstein *et al.* (9) published a modification of a method using tributyrin introduced by them in 1943 (10). This method was a welcome one because of the 1-hour incubation period as opposed to the 24 hours required with the use of olive oil and was rather widely accepted in spite of the fact that there was no clinical evidence for its validity as a diagnostic test. A number of European investigators, including Lagerlöf (11, other references cited therein), employed the stalagmometric technique with tributyrin as substrate and obtained increased values in acute pancreatitis when calcium oleate was added to the reaction.

[*] Based on the method of Cherry and Crandall (4).
[†] See text for definition.

86

It has now been shown that "tributyrinase" as measured by the technique of Goldstein *et al.* does not increase in cases of acute pancreatitis and that the hydrolysis of tributyrin that does occur with serum is due to a pseudocholinesterase (1, 12, 13). Admittedly, the olive oil technique as proposed by Cherry and Crandall is not an ideal one, notably because of the relatively slow rate of hydrolysis, the none-too-sharp end point obtained in titration, and the difficulty in preparing a stable emulsion for substrate. A study (2) of this method has revealed the striking specificity of the lipolytic enzyme occurring in serums from cases of acute pancreatitis. Since this enzyme differs considerably in behavior from the lipolytic enzymes in pancreatic extracts, the great amount of data obtained on such extracts cannot be freely applied to any consideration of the problem of the determination of that lipolytic enzyme appearing in serums from cases of acute pancreatitis. To avoid confusion in terminology, the term "pancreatitis-lipase" is suggested. It is clear that only active serums from patients with this disease should be used in investigative work concerned with "pancreatitis-lipase."

Seligman and Nachlas (14) have proposed a method employing β-naphthyl laurate as substrate, and it appears that increased values may occur in pancreatitis. Since this has been questioned (15) and relatively little data are as yet available (16), final decision on the validity of this method must be delayed. Thus, in spite of its shortcomings, it appears that the only method which can be recommended at this time is that of Cherry and Crandall, which is presented below with some modification.

Reagents

1. Olive oil substrate. To 1 part olive oil, U.S.P. grade or better (Old Monk Olive Oil, Old Monk Co., 718 North Aberdeen, Chicago, Illinois, is excellent), add 1 part 5% gum acacia containing 0.2% sodium benzoate. Emulsify either (a) by passing through a hand homogenizer repeatedly until a smooth white emulsion is obtained (usually 5–12 times required) or (b) by mixing in a Waring Blendor for 15 minutes. Prepare an estimated 4 weeks' supply and store in refrigerator. Discard if olive oil separates out.

2. M/15 phosphate buffer, pH 7.0. Make 0.64 g. Na_2HPO_4 (anhydrous) + 0.36 g. KH_2PO_4 to 100 ml. with water.

3. 0.050 N NaOH.

4. Thymolphthalein indicator. 1% in denatured alcohol.

5. Denatured alcohol. Use formula 3A, anhydrous, which is 95% ethanol and 5% methanol.

NOTE: Denatured alcohol is inexpensive and tax exempt. Absolute or 95% ethyl alcohol is suitable.

Procedure

To a test tube add 3.0 ml. water, 0.50 ml. buffer, 2.0 ml. substrate and 1.0 ml. serum. Mix and stopper.

NOTE: By employing a microburet, the test can easily be scaled down fourfold. The test can thus be run on 0.25 ml. of serum. Do *not* run the test with full amounts of reagents and reduced amount of serum and then multiply the results by a factor to correct for reduced sample size as the results so obtained are not equivalent (2).

Incubate overnight (about 16 hours) at 37°C.

NOTE: The original Cherry-Crandall technique called for a 24-hour incubation period. The shape of the curve relating units of activity and time of incubation deviates sufficiently from zero order that there is relatively little increase after 16 hours. Practically, it is more feasible to set up, in late afternoon, all samples received during the day and titrate them the next morning. A 4-hour incubation time has been suggested (1) and the results are approximately one-half those obtained with 16–24 hours' incubation (1, 2). If specimens are received in the morning the 4-hour test is advantageous, but if they are received in late afternoon it is not. Use of both periods of incubation would lead to confusion.

Serum samples are stable up to 1 week at room (25° C.) temperature and for many weeks at refrigerator temperature (2).

Remix; pour into a 50-ml. Erlenmeyer flask. Add 3.0 ml. denatured alcohol to the test tube, shake vigorously, and add to the flask.

Add 2 drops of indicator and titrate with 0.050 N NaOH to a distinct blue color (color fades).

NOTE: The equivalence point of titration of the free fatty acids produced by hydrolysis, as determined by titration curves (2), is at about pH 10.5. The original Cherry-Crandall method employed phenolphthalein as indicator, which, under the conditions of the titration, changes color at about pH 8.8. At this pH only about 70% of the fatty acids are titrated. Titration to pH 10.5 with a pH meter would be the most accurate technique but also the most laborious. Thymolphthalein changes to a distinct blue at about pH 10.5 under the conditions given. The end point is not very sharp, becoming a pale blue at pH 10.1, a distinct blue at pH 10.5, and a deep blue at pH 10.8. It is estimated, however, that an error of less than 5% of the total titration is involved in titrating to a "distinct" blue. With a highly icteric serum the end point is a green color.

A blank should be run with each set of determinations. There are two alternatives: (a) To a test tube add 3.0 ml. water, 0.50 ml. buffer, and 1 ml. serum. Place in boiling water for 5–10 minutes. Add 2.0 ml. substrate and proceed as for unknown. (b) To a test tube add 3.0 ml. water, 0.50 ml. buffer, and 2.0 ml. substrate. Mix and stopper. Proceed as for an unknown but immediately prior to titration add 1.0 ml. serum.

NOTE: The blank titration includes the alkali required by the buffer, emulsion and serum to bring the pH to the end point used. This remains practically constant for any particular olive oil and the purpose of running a blank with each set of determinations is to check on technique and reagents. It is perhaps not necessary to run a blank for each serum and, in fact, if one unknown is run and the quantity of sample is limited, a blank employing another serum might be used.

Calculation

Lipase units = Milliliters of 0.050 N NaOH required for unknown

− milliliters of 0.050 N NaOH required for blank.

Precision of Method

The reproducibility of the method at a level of 3.5 units is about ±10% (95% confidence limits) (2).

NOTE: The method in the checker's laboratory was independently performed by two technicians. Their results were within 0.3 of a lipase unit in ten analyses, within 0.2 of a unit in eight analyses, and within 0.1 of a unit in seven analyses. Samples of serum analyzed in the laboratory of the editor (D. S.) by an independently modified Cherry-Crandall method and sent to the checker showed the following results:

No.	Editor	Checker
14	1.7	2.2, 2.1
21	1.5	1.3, 1.0
26	2.7	3.2, 3.0
55	3.2	3.4, 3.2

Normal Values

The serum of healthy adults contains up to 1.5 lipase units (95% limits). Elevated values of "pancreatitis lipase" obtained by the use of olive oil substrate have been reported in acute pancreatitis (1, 2, 5–8), acute attacks of chronic relapsing pancreatitis (18), carcinoma of the pancreas (8, 19), cirrhosis and hepatitis (6, 7), biliary tract disease and duodenal ulcer with penetration into the pancreas (8).

NOTE: It has been reported (17) and confirmed (2) that hemoglobin inhibits the lipolytic activity. A normal result from a hemolyzed sample is, therefore, subject to question, and an elevated result, while significant, is lower than its true value When possible, another sample should be obtained.

REFERENCES

1. Bunch, L. D., and Emerson, R. L., Serum lipase determination. Four-hour technique with olive oil substrate. *Clin. Chem.* **2**, 75–82 (1956).
2. Henry, R. J., Sobel, C., and Berkman, S., On the determination of "pancreatitis-lipase" in serum. *Clin. Chem.* **3**, 77–89 (1957).
3. Rona, P., and Michaelis, L., Uber Ester- und Fettspattung in Blute und im Serum. *Biochem. Z.* **31**, 345–354 (1911).
4. Cherry, I. S., and Crandall, L. A., Jr., Specificity of pancreatic lipase: Its appearance in blood after pancreatic injury. *Am. J. Physiol.* **100**, 266–273 (1932).
5. Comfort, M. W., and Osterberg, A. E., Lipase and esterase in blood serum. Their diagnostic value in pancreatic disease. *J. Lab. Clin. Med.* **20**, 271–278 (1934).
6. Comfort, M. W., Serum lipase. Its diagnostic value. *Am. J. Digest. Diseases and Nutrition* **3**, 817–821 (1936).
7. Comfort, M. W., and Osterberg, A. E., The value of determination of the concentration of serum lipase and serum amylase in the diagnosis of disease of the pancreas. *Proc. Staff Meetings Mayo Clinic* **15**, 427–432 (1940).
8. Johnson, T. A., and Bockus, H. L., Diagnostic significance of determinations of serum lipase. *Arch. Internal Med.* **66**, 62–77 (1940).
9. Goldstein, N. P., Epstein, J. H., and Roe, J. H., Studies of pancreatic function. IV. A simplified method for the determination of serum lipase, using aqueous tributyrin as substrate, with 100 normal values by this method. *J. Lab. Clin. Med.* **33**, 1047–1051 (1948).
10. Goldstein, N. P., and Roe, J. H., Studies of pancreatic function. I. The determination of the lipidolytic enzymes of blood serum. *J. Lab. Clin. Med.* **28**, 1368–1379 (1943).
11. Lagerlöf, H. O., Normal serum esterase and pancreatic lipase in diseases of biliary ducts and pancreas. *Acta Med. Scand.* **128** Suppl. 196, 399–410 (1947).
12. Richter, D., and Croft, P. G., Blood esterases. *Biochem. J.* **36**, 746–757 (1942).
13. Mendel, B., and Rudney, H., Studies on cholinesterase. I. Cholinesterase and pseudocholinesterase. *Biochem. J.* **37**, 59–63 (1943).
14. Seligman, A. M., and Nachlas, M. M., The colorimetric determination of lipase and esterase in human serum. *J. Clin. Invest.* **29**, 31–36 (1950).
15. Delcourt, A. A., Rubin, C. E., Palmer, W. L., and Kirsner, J. B., Evaluation of a new technique for lipase determination. *J. Lab. Clin. Med.* **42**, 310–315 (1953).
16. Seligman, A. M., Glotzer, P., and Persky, L., Preliminary clinical observations of a serum lipase activity as determined by a colorimetric method. *Surgery* **30**, 923–930 (1951).
17. Tauber, H., New olive oil emulsion for lipase and new observations concerning "serum lipase." *Proc. Soc. Exptl. Biol. Med.* **90**, 375–378 (1955).
18. Farrar, J. T., Pancreatitis, medical aspects. *Med. Clin. N. Am.* pp. 1393–1401 (1954).
19. Ingelfinger, F. J., Diagnosis of cancer of the pancreas. *New Engl. J. Med.* **235**, 653–661 (1946).

NITROGEN BY THE KJELDAHL METHOD*

Submitted by: REGINALD M. ARCHIBALD, Hospital of the Rockefeller Institute for Medical Research, New York, New York

Checked by: ELIZABETH G. FRAME, Clinical Chemistry Service, Clinical Pathology Department, National Institutes of Health, Bethesda, Maryland
DOROTHY SENESKY, Division of Biochemistry, Graduate Hospital of the University of Pennsylvania, Philadelphia, Pennsylvania

Introduction

In general, simpler and more rapid methods than Kjeldahl determination will be used to measure serum, plasma (1, 2), or urine (3) proteins. However, the Kjeldahl method is the standard on which the other methods are based. Hence, occasion will arise for the chemist to check the accuracy of these other methods against this standard. Normal and pathological serums of which the total nitrogen and nonprotein nitrogen (NPN) have been determined by Kjeldahl measurement serve as the secondary standards for other methods.

Principle

Organic matter of the sample to be analyzed is oxidized by heating it with sulfuric acid in the presence of a catalyst. All combined nitrogen in the sample is thereby converted to ammonium sulfate nitrogen. An excess of alkali is then added and the ammonia which is thus liberated is distilled into acid solution. In the macro-Kjeldahl procedure this ammonia is trapped by a large excess of boric acid and then measured by direct titration with standardized diluted strong acid. In the micro-Kjeldahl procedure the ammonia is distilled into an accurately measured small excess of standard H_2SO_4. The acid not neutralized by the ammonia is then titrated with standard NaOH.

In the clinical work, amino nitrogen, imino nitrogen, amide nitrogen, and ammonia nitrogen usually include all of the combined nitrogen, and the standard procedure outlined will convert to NH_3

* Based on the method of Hiller, Plazin, and Van Slyke (11).

nitrogen all that is not already in that form. When, however, some
or all of the combined nitrogen is in a more oxidized state, as in
nitrate, nitrite, or nitro derivatives, preliminary treatment with
phenolic compounds such as 1-naphthol and pyrogallol and subsequent reduction of such acceptors with sodium thiosulfate make
possible more complete measurement of nitrogen originally present
in an oxidized form (4, 5). The absence of nitrate, nitrite, and nitro
groups from clinical specimens should not be assumed. Nitrogen in
these forms is present in small amounts more often than has been
generally realized (6).

I. MACRO-KJELDAHL METHOD

Reagents

1. *Powdered potassium sulfate, K_2SO_4 (ammonia-free).*
2. *Concentrated sulfuric acid, c. p. H_2SO_4, sp. gr. 1.84.*
3. *Mercuric sulfate solution.* Add 12 ml. of concentrated H_2SO_4
to water and make up to 100 ml. with water. Dissolve 10 g. of red
mercuric oxide in this solution.
4. *Zinc dust (not granulated zinc), ammonia-free.*
5. *Concentrated sodium hydroxide solution (approximately 18 N
NaOH).*
6. *Boric acid, 4% solution.* Dissolve 40 g. of H_3BO_3 in water and
make up to 1 l. Boil to remove CO_2 then store in a Pyrex container
(7, 8) to minimize solution of titratable alkali from the glass.
7. *Sulfuric acid $N/14.01$ solution, (0.07139 N).*
8. *Indicator.* Bromocresol green, 0.1% solution, in 95% ethyl
alcohol; or methyl red, 0.1% solution, in 95% ethyl alcohol. (8)

Procedure

Into a dry 500-ml. Kjeldahl flask, measure the sample (which
contains 15–90 mg., of nitrogen; e.g. 5 ml. serum) to be analyzed.
Add 10 g. of potassium sulfate (see notes 1 and 2), 10 ml. of the mercuric sulfate solution (see note 3), and 20 ml. of concentrated sulfuric acid (see notes 4 and 5).

NOTE 1: Potassium acid sulfate is formed. This is hydrated by water formed during
oxidation of the organic compounds. When it is heated it loses water much more
readily than do the hydrates of sulfuric acid (9). Thus potassium sulfate serves to
raise the boiling point of the mixture and to hasten digestion.

NOTE 2: If Kjeldahl determinations are to be run frequently, it is convenient to make glass spoons (10) which will measure with sufficient accuracy and reproducibility the desired amount of K_2SO_4 and of zinc dust.

NOTE 3: The mercuric salt serves as a catalyst and has been demonstrated to yield appreciably higher results than when selenium or mixtures of selenium and copper are used (11, 12).

NOTE 4: Care should be taken to avoid accidental contamination of the digest with ammonium salt. Even though the contaminant may not be visible as a "frost," ammonium salts tend to accumulate especially where nonvolatile acid is in contact with room air. Thus, openings in the lead or glass pipes which are often used to draw off fumes from the mouths of Kjeldahl flasks during digestion with acid are likely to be coated with condensed H_2SO_4, hence also with $(NH_4)_2SO_4$. Such openings should be wiped with a clean moist towel prior to each use. Similarly, contamination from the mouths of acid reagent bottles should be avoided. Obviously, room air should not be contaminated with ammonia during any stage of the procedure.

NOTE 5: In those instances when nitrates, nitrites, or nitro derivatives make up at least part of the fixed nitrogen of the sample, the digestion procedure up to this point is modified as follows (4, 5): To the dry sample in the Kjeldahl flask add approximately 1 g. of a mixture of equal parts of 1-naphthol and pyrogallol. Then add 20 ml. concentrated H_2SO_4. Heat the flask and contents on a steam bath for 30 minutes or until the sample is completely dissolved. Then add 5 g. of sodium thiosulfate, $Na_2S_2O_3$. After another 30 minutes, heat the flask on a digestion rack until the mixture carbonizes. Then cool the flask and add 10 g. K_2SO_4 and 10 ml. of mercuric sulfate solution and continue the digestion.

With the neck of the flask inclined 30°–60° from horizontal, digest over low heat until frothing ceases and nearly all of the water has been driven off. If a flame is used, no part of it should touch the flask above the level of the liquid. This digestion with heat is continued for 2 hours after the mixture has become clear. A salt cake will form as the flask and contents are cooled to room temperature. From a 200-ml. portion of water add about 10 ml. to the salt cake. Swirl the contents of the flask gently until the salt is dissolved. Then add the remainder of the 200 ml. of water. Cool again and have a still with a glass trap (see note 6) and a recently rinsed glass condenser tube (see note 7) ready for use.

NOTE 6: Because the excess of NaOH added to the digest is very large compared with the amount of ammonia to be measured, it is important that no spray pass from the Kjeldahl flask into the condenser. A good trap minimizes danger of such an error (13).

NOTE 7: In many older models of Kjeldahl stills the condenser tubes are made of tin. Although these are very satisfactory when catalysts other than mercuric salts are used, they would eventually be destroyed by amalgamation with the small amounts of mercury which distill over after reduction of the catalyst with zinc (14).

Place 50 ml. of the 4% boric acid solution preferably in a 500-ml. wide-mouth Erlenmeyer flask. The lower tip of the condenser tube or a delivery tube attached thereto (which has been coated *very* lightly with silicone grease) should be just beneath the upper level of the boric acid solution. Tilt the Kjeldahl flask and allow 50 ml. of 18 N NaOH to run down its neck to form a layer beneath the diluted digest. Then add 2 g. of zinc dust (see notes 2 and 8).

Note 8: When alkali is added to a solution which contains both ammonium and mercuric ions, a considerable portion of the ammonia combines with the mercuric salt to form a complex from which the ammonia cannot be recovered by distillation. To prevent such loss, zinc dust is added, with the result that metallic mercury is formed which cannot bind the ammonia. The presence of zinc serves also to liberate a stream of small bubbles of hydrogen which promote a smoothly boiling distillation. Thereby the danger of bumping is minimized.

Connect the mouth of the Kjeldahl flask to the trap above the condenser and swirl the contents until mixed. Apply heat so that ammonia and water distill into the receiving flask. Distillation continues until about 200 ml. of distillate are collected or until bumping begins. About 1 minute before distillation is discontinued, lower the receiving flask so that the last 2 ml. of distillate will rinse off the tip of the condenser into the distillate. Add 5 drops of indicator solution (see note 9) to the distillate.

Note 9: Meeker and Wagner (7, 8) recommend the use of a mixture of methyl red and methylene blue but seem to prefer the methyl red alone. They state that titration with methyl red alone and methyl red plus methylene blue are equally accurate but that the single indicator gives a sharper end point. The mixed indicator has the advantage that it signals earlier this approach to the end point.

Titrate the distilled ammonia with $N/14.01$ sulfuric acid until the color matches that in the control flask. Prepare this control by measuring into a flask of the type used as a receiver 50 ml. of boric acid, 5 drops of the same indicator, and enough boiled distilled water to make the liquid volume equal to that in the receiver which contains the distillate.

Simultaneously, blank analyses are obtained by performing all the steps in the same order and by use of quantities of each reagent equal to those used with the unknowns (except that no unknown sample is added to the digestion flask used for the blanks).

NOTE 10: If the K_2SO_4, H_2SO_4, or catalyst contain a measurable amount of nitrate or nitrite, it is wise to add to the blank an amount of nitrogen-free cane sugar equal in weight to the unknown sample used in the other flasks. This will reduce the nitrates or nitrites in the reagents about as effectively as will the unknown [though not as completely as would treatment with phenols and thiosulfate (see note 5)]. The purity of reagents now available in most countries is such that addition of sugar to the blank is seldom necessary.

Calculations

Milligrams of nitrogen in sample analyzed (see note 11) equal the milliliters of standard acid used in titration of distillate minus milliliters used in titration of blank.

NOTE 11: Milligrams of serum protein in the sample digested equal milligrams of nitrogen multiplied by 6.25. When the sample is made up almost entirely of gamma globulin, the factor 6.24 is more accurate. Similarly for albumin, 6.27 is preferred (11).

Urine Analyses

For total nitrogen of urine a 5.00-ml. aliquot (sp. gr. 1.020 or less) is treated as described for the 5 ml. of serum. If total NPN of a proteinaceous urine is to be determined, a 5-ml. aliquot of urine (sp. gr. 1.020 or less) is treated with 1.0 ml. 10% solution of sodium tungstate (see note 12) and 1.0 ml. of $\frac{2}{3}$ N H_2SO_4 and centrifuged. A few extra drops of $\frac{2}{3}$ N H_2SO_4 may be required to yield a clear supernatant fluid. A 5.00-ml. aliquot of clear fluid is then used for the digestion.

NOTE 12: It should not be assumed that trichloroacetic acid (TCA) filtrates of urine and blood contain no protein. Urine contains a TCA-soluble mucoprotein. This is present in relatively large amounts in urine of nephrotic individuals (15). A TCA-soluble protein fraction has been found also in peanuts and cottonseed (16). TCA filtrates of plasma or serum are believed to contain relatively little protein, but the content of TCA filtrates of whole blood is less certain. If, nevertheless, the NPN of a TCA filtrate is desired, heat a mixture of 20 ml. urine (sp. gr. 1.020 or less), 20 ml. 10% TCA, and 0.5 g. NaCl in a boiling water bath for 3 minutes, then filter or centrifuge. Take a 10-ml. aliquot for digestion. As nitrogenous lipids are removed by TCA and tungstate, the filtrates do not contain all the .NPN. If, in the NPN figure, one wishes to include the lipid nitrogen, precipitation of protein with organic solvent mixtures (alcohol-ether-acetone) is advisable.

Total Nitrogen of Feces

Feces should be collected in dilute acid (e.g. 1% boric acid solution) to prevent loss of ammonia. (Drying of feces results in loss of ammonia.) They are blended (e.g. in a Waring Blendor) so that a representative aliquot may be obtained. Alternatively they can be suspended in concentrated H_2SO_4, mixed, and sampled. An aliquot which contains 20–30 mg. of nitrogen ($\frac{1}{50}$ of the daily output of an adult) is transferred to the Kjeldahl flask and digested.

II. MICRO-KJELDAHL METHOD

Reagents

1. *K_2SO_4*
2. *$HgSO_4$ solution*
3. *Concentrated H_2SO_4* } as described for the macro-Kjeldahl method.
4. *Zinc dust*

5. *Sodium hydroxide, approximately 10 N.* Dissolve 400 g. of NaOH in water and dilute to 1 l.

6. *Standard 71.4 millimolar ammonium chloride solution.* Used for checking the micro-Kjeldahl distillation procedure. Dissolve 0.3820 g. of NH_4Cl, "analytical reagent" grade, in water and dilute to 100 ml. One milliliter contains 1 mg. of nitrogen.

7. *Acetate buffer, 0.2 M, pH 5.* Dissolve 27.22 g. of $NaC_2H_3O_2$. $3H_2O$ in water and make to 1 l. Add 427 ml. of 0.2 N acetic acid, (standardized by titration against 0.1 N NaOH with phenolphthalein as an indicator).

8. *Alizarin red solution, 0.1% in water.*

9. *0.01428 N (N/70) H_2SO_4.* Prepare by dilution of 14.28 ml. of 1 N acid to 1 l. with water.

10. *0.01428 N (N/70) NaOH solution.* Store in a plastic container or a paraffin-lined bottle. Protect against CO_2 by a soda lime tube. Standardize the solution daily by titration against 10-ml. portions of the 0.01428 N H_2SO_4, with the same pH and volume at the end point as is described below for titration of distilled ammonia.

11. *Boiling chips.* Use a piece of Teflon about 2 mm. x 2 mm. x 6 mm. cut from a Teflon sheet or Norton's Alundum Chips, No. 14, black.

Procedure

Into a Pyrex glass tube 22–25 mm. x 200 mm., measure 0.5 g. of K_2SO_4 (see notes 13 and 14) and the sample of unknown which contains 0.2–2.0 mg. of nitrogen (e.g. 0.1 ml. of serum).

NOTE 13: It is convenient to add the K_2SO_4 (see note 2) to the bottom of the digestion tube with the aid of a glass funnel which has a 12-cm. stem, 1 cm. in diameter.

NOTE 14: The items in 1–4, 6–8, 10, and 11 under the macroprocedure apply also to the micromethod. Traces of NH_3 in the room air may cause even greater percentage error in the microprocedure than in the macromethod. In this connection it is noteworthy that each lighted cigarette can liberate at least 1 mg. of NH_3 nitrogen per minute.

Then add 0.5 ml. of the mercuric sulfate solution, 1.0 ml. of concentrated sulfuric acid, and 1 piece of Teflon or 3 pieces of alundum. Boil mixture gently until the water is boiled off. Then adjust the heat so that the concentrated digest boils with very slight motion. This digestion continues for 30 minutes after the mixture has become entirely clear. About 2 minutes after completion of the digestion, but before the contents solidify, wash down the sides of the tube with 3 ml. of water. Grease the lip of the digestion tube lightly with silicone grease to avoid loss during the quantitative transfer to the distillation apparatus. Steam out the still for 30 minutes before each series of distillations.

NOTE 15: If the distillation apparatus described by Parnas and Wagner (17) is used, it should be modified so that the inner tube of the distilling flask is at the lowest position when the flask is in position for distillation, and the flask should have a conical rather than a round bottom. Such stills are available from Machlett & Sons, 220 East 23rd Street, New York.

Transfer the contents of the digestion tube into the distillation flask with four portions of water, approximately 2 ml. each. Add 0.2 g. of zinc dust (see note 2) to the third washing in the funnel of the distillation apparatus. Admit the mixture to the flask and follow with the fourth washing. Deliver 4 ml. of $10\ N$ NaOH into the distillation flask and distill into 10.00 ml. of $0.01428\ N$ H_2SO_4 , with the tip of the condenser below the surface of the acid. Distill enough liquid to insure quantitative transfer of the ammonia to the standard acid. The volume of liquid which must be so transferred by distillation will vary from one apparatus to another and should be predetermined for each distillation apparatus by trial runs with the standard 71.4 millimolar ammonium chloride solution. Then continue the distillation for 1 minute after the receiving flask has been lowered so that the tip of the condenser tube is above the surface of the standard acid. To the distillate add 0.8 ml. of 0.1% alizarin red indicator solution and titrate with 0.01428 N NaOH from a 10-ml. buret until the color matches that of acetate buffer solution in a control flask. The volume of liquid at the end of the titration should be nearly equal to that in the control flask. To prepare the control, measure into a 125-ml. Erlenmeyer flask 7 ml. of 0.2 N acetate buffer, 63 ml of distilled water, and 0.8 ml. of 0.1% alizarin red solution. A new control should be made up every 3 days or more often if mold growth becomes visible therein. Blank analyses are run through the entire procedure.

Calculations

Milligrams nitrogen in sample analyzed (see note 11) equals (milliliters of 0.01428 N NaOH required to back-titrate the blank minus milliliters required to titrate distillate of unknown) multiplied by 0.2.

NPN of Blood (*See Note 12*)

A 20.0-ml. aliquot of a 1:10 tungstic acid filtrate of blood or of plasma is taken for digestion.

Total Nitrogen of Spinal Fluid

A 2.00-ml. aliquot is used for digestion.

Total Nitrogen of Urine

A 0.100-ml. aliquot of urine is taken for digestion.

Total Nitrogen in Feces and Other Samples

Representative figures for tissues or foods can be obtained by pretreating a large weighed aliquot with concentrated H_2SO_4 in the same manner as was described for feces. An aliquot which contains 0.2–2.0 mg. of nitrogen is taken for digestion. This corresponds to about $\frac{1}{500}$ of the daily output of feces.

REFERENCES

1. Reinhold, J. G., Total protein, albumin, and globulin. *In* "Standard Methods of Clinical Chemistry" (M. Reiner, ed.), Vol. 1, pp. 88–97. Academic Press, New York, 1953.
2. Van Slyke, D. D., Hiller, A., Phillips, R. A., Hamilton, P. B., Dole, V. P., Archibald, R. M., and Eder, H. A., The estimation of plasma protein concentration from plasma specific gravity. *J. Biol. Chem.* **183**, 331–347 (1950).
3. Hiller, A., Greif, R. L., and Beckman, W. W., Determination of protein in urine by the biuret method. *J. Biol. Chem.* **176**, 1421–1429 (1948).

4. Bradstreet, R. B., Kjeldahl method for organic nitrogen. *Anal. Chem.* **26**, 185–187 (1954).

5. Bradstreet, R. B., Determination of nitro nitrogen by the Kjeldahl method. *Anal. Chem.* **26**, 235–236 (1954).

6. Schaus, R., Griess' nitrite test in diagnosis of urinary infection. *J. Am. Med. Assoc.* **161**, 528–529 (1956).

7. Meeker, E. W., and Wagner, E. C., Titration of ammonia in presence of boric acid. *Ind. Eng. Chem., Anal. Ed.* **5**, 396–398 (1933).

8. Wagner, E. C., Titration of ammonia in presence of boric acid in the macro-, semimicro- and micro-Kjeldahl procedures with methyl red indicator and the color-matching end point. *Ind. Eng. Chem., Anal. Ed.* **12**, 771–772 (1940).

9. Gunning, J. W., Ueber eine Modification der Kjeldahl-Methode, *Z. anal. Chem.* **28**, 188–191 (1889).

10. Van Slyke, D. D., Hiller, A., Weisiger, J. R., and Cruz, W. O., Determination of carbon monoxide in blood and of total and active hemoglobin by carbon monoxide capacity. Inactive hemoglobin and methemoglobin contents of normal human blood. *J. Biol. Chem.* **166**, 121–148 (1946).

11. Hiller, A., Plazin, J., and Van Slyke, D. D., A study of conditions for Kjeldahl determination of nitrogen in proteins. Description of methods with mercury as catalyst, and titrimetric and gasometric measurements of the ammonia formed. *J. Biol. Chem.* **176**, 1401–1420 (1948).

12. Miller, L., and Houghton, J. A., The micro-Kjeldahl determination of the nitrogen content of amino acids and proteins. *J. Biol. Chem.* **159**, 373–383 (1945).

13. Peters, J. P., and Van Slyke, D. D., Total and non-protein nitrogen. "Quantitative clinical chemistry," Vol. 2, p. 521, Williams & Wilkins, Baltimore, Maryland, 1932.

14. Andersen, A. C., and Jensen, B. N., Zur Bestimmung des Stickstoffs nach Kjeldahl. *Z. anal. Chem.* **67**, 427–448 (1926).

15. Beckman, W. W., Hiller, A., Shedlovsky, T., and Archibald, R. M., The occurrence in urine of a protein soluble in trichloroacetic acid. *J. Biol. Chem.* **148**, 247–248 (1943).

16. Fontaine, T D., Irving, G. W. Jr., and Markley, K. S., Peptization of peanut and cottonseed proteins. *Ind. Eng. Chem.* **38**, 658–662 (1946).

17. Parnas, J. K., and Wagner, R., Uber die Ausführung von Bestimmungen kleiner Stickstoffmenger nach Kjeldahl. *Biochem. Z.* **125**, 253–256 (1921).

NONPROTEIN NITROGEN*†

Submitted by: ELIOT F. BEACH, Biochemical Laboratory, Metropolitan Life
Insurance Company, New York, New York
Checked by: RUTH D. MCNAIR, Department of Pathology, Providence Hospital,
Detroit, Michigan
PHILIP G. ACKERMANN, Saint Louis Chronic Hospital; and Division
of Gerontology, Washington University School of Medicine, Saint
Louis, Missouri
JEAN MARINO, Division of Biochemistry, Graduate Hospital of the
University of Pennsylvania, Philadelphia, Pennsylvania

Introduction and Principles

Nonprotein nitrogen is determined by conversion to ammonia
by micro-Kjeldahl digestion of the protein-free filtrate of blood.
Ammonia is then estimated colorimetrically using Nessler's reagent.
This is the classic method of Folin and Wu (1). Numerous modifications have appeared. Some have introduced new types of catalysts for the Kjeldahl digestion, others describe different ways of
preparing Nessler's solution.

Reagents

1. 10% sodium tungstate $(Na_2WO_4.2H_2O)$.

2. 0.667 (2/3) N sulfuric acid.

3. Regular acid digestion mixture. Mix 300 ml. of 85% phosphoric
acid with 50 ml. of 5% copper sulfate solution. Add to this 100 ml.
of concentrated sulfuric acid. Store tightly covered to prevent absorption of ammonia fumes and water.

4. Dilute acid digestion mixture. Dilute the regular acid digestion
mixture described above with an equal volume of water.

* Based on the method of Folin and Wu (1).
† Acknowledgment is made of the technical assistance of Mr. Gelson Toro,
Washington University School of Medicine, St. Louis, Missouri, Mr. Joseph
Dworkin, Metropolitan Life Insurance Company, New York City, and the medical
technologists of Providence Hospital, Detroit, Michigan.

5. Nessler's solution. Place 150 g. of potassium iodide in a 500-ml. flask and add 100 g. of iodine followed by 100 ml. of water and 140–150 g. of metallic mercury. Shake the contents vigorously and continuously for 7–15 minutes until the dissolved iodine has nearly disappeared. The solution becomes quite hot. When the iodine color becomes visibly pale, cool the flask under running water and continue to shake until the red color of the iodine is completely replaced by the green color of the double iodide of mercury and potassium ($HgI_2.2KI$). Decant the solution from the excess mercury and wash with several portions of water, combining solution and washings. Dilute the solution to 2 l.

The final Nessler's solution can be prepared from the potassium mercuric iodide as follows: Introduce into a large bottle 3500 ml. of 10% sodium hydroxide; add 750 ml. of the double iodide solution and 750 ml. of water and mix. A small amount of precipitate may form in the Nessler's solution on standing. If this occurs, only the clear supernatant solution should be used. The 10% sodium hydroxide must be quite accurately standardized to secure good results. It can be prepared from a strong solution of NaOH (containing 75 g./100 ml.) which has been allowed to stand until the carbonate has settled, the clear supernatant being used. The alkalinity of the Nessler's solution is important and should be checked. Titration of the reagent with standard acid using phenolphthalein should indicate a normality close to 1.75 (1.74–1.82). A further check on the reagent should be made by titrating 1 ml. of the diluted digestion mixture with the Nessler's solution. This titration should require 9.0–9.3 ml. of Nessler's solution.

NOTE: It is frequently useful for small laboratories to prepare Nessler's solution directly from the double salt of mercury and potassium iodide. The reagent may also be prepared according to Vanselow (2) from mercuric iodide and potassium iodide. This is available commercially in satisfactory form.

6. Standard ammonium sulfate. Dissolve 0.472 g. of reagent grade ammonium sulfate (previously dried overnight in an oven at 100° C.) in water and dilute to 2000 ml. after adding a few drops of concentrated sulfuric acid. This solution is stable and contains 0.0500 mg. of nitrogen per milliliter.

Procedure

Transfer to an appropriate-sized flask exactly 7 times as much water as you plan to use of blood. Follow this with 1 volume of whole bood. Mix. Add 1 volume of 10% sodium tungstate solution and mix. Then add, with continuous shaking, 1 volume of 0.667 N sulfuric acid. Stopper and shake the flask. Practically no foam will form if all of the protein has been precipitated. Allow to stand 10 minutes; during this time the mixture changes to a chocolate-brown color. Pour the mixture into a funnel containing a dry filter paper. Cover the funnel with a watch glass to minimize evaporation and collect the filtrate in a clean, dry flask. If the protein precipitation is correctly done, the filtrate is perfectly clear. This condition must be met to secure accurate nonprotein nitrogen values on the filtrate.

NOTE: A number of methods for deproteinizing blood may be used in lieu of the tungstic acid method described here. Ten per cent trichloroacetic acid and metaphosphoric acid filtrates are equally satisfactory, as demonstrated by Hiller and Van Slyke (3). The Somogyi zinc precipitation method removes glutathione, ergothionine, uric acid, and part of the creatinine and yields NPN values about 10 mg./100 ml. lower than do the other methods (4).

Place 5.00 ml. of the protein-free filtrate in a large, dry Pyrex test tube (200 mm. x 25 mm.) graduated at 35 and 50 ml. Add 1.0 ml. of dilute acid mixture and two small perforated beads or quartz pebbles to aid in preventing bumping. Place the test tube vertically in a clamp attached to a small ring stand and boil the contents vigorously to evaporate the water. Vigorous heating will aid in preventing bumping. Boil in this way until heavy white fumes of SO_3 begin to fill the tube. Remove the flame and cover the mouth of the tube with a large glass marble or with a small watch glass and continue to heat with a reduced flame for 2 minutes with the acid mixture barely boiling. Care should be exercised to prevent appreciable escape of SO_3 fumes, and boiling must be kept at a very minimum to prevent troublesome cloudiness of the solutions later on. Oxidation is almost always apparently complete by 2 minutes, but if this is not the case further heating is required. Allow the contents of the tube to cool for 70–90 seconds. Add water to bring the volume to within a few millimeters of the 35-ml. mark and allow the tubes to cool to room temperature. Cooling at this stage is most important because nesslerization of even slightly warm solutions frequently causes mild turbidity and erratically high results. During this period prepare a standard and blank so that standard, blank, and unknown may be nesslerized at approximately the same time. The standard is prepared as follows: Place 3.00 ml. of standard ammonium sulfate solution (containing 0.150 mg. of ammonia nitrogen) in a 50-ml. tube similar to that used for unknowns, add 1 ml. of digestion mixture, and dilute to nearly 35 ml. The blank is similarly made, omitting the ammonium sulfate solution. It is desirable to digest the standard and blank in order to provide the same conditions as those provided for the unknown.

When all solutions—standard, blank, and cooled unknowns—are ready, add, with continuous mixing, 15 ml. of Nessler's solution from a graduated cylinder. In adding Nessler's solution the contents of the tube should be continuously and gently rotated to secure, rapid, complete mixing of the materials. Uneven alkalinization of the solution on adding the reagent may cause turbidity to develop. Dilute the unknown to the 50-ml. mark immediately upon completing the addition of the Nessler's solution with water and mix by inversion. Likewise with the standard, dilute to the 50-ml. mark with water immediately after adding the Nessler's solution and mix. Allow the solutions to stand 10 minutes and read within the next 10 minutes. Long standing frequently results in turbidity. Occasionally there may be turbidity in the solutions before Nessler's solution is added. This is due to some gelatinous silica-like material produced during digestion. It may be removed by centrifugation after nesslerizing or by allowing it to settle by gravity. Turbidities developing after the addition of Nessler's solution render the determination valueless.

NOTE: In one of the laboratories testing this method, all samples were centrifuged after nesslerization routinely to remove this material. The other two testing laboratories found less of it and omitted centrifuging. We believe this difference may arise from more or less vigorous digestion. The individual worker is advised to decide whether in his hand the unknowns appear to need centrifugation.

Turbidity will often develop in solutions with high NPN values. Furthermore, excessive NPN levels will be out of the range of the photometer and will deviate greatly from Beer's law. This will call for a proper dilution of the high unknown and a redetermination of the NPN.

Formation of a red precipitate on nesslerization of solutions, with consequent failure to develop the characteristic color of amber yellow, usually indicates that the alkali of the Nessler's solution and the acid concentration of the digestion mixture are not correctly balanced.

For photometric measurement, set the photometer to zero absorbance with the blank and read the absorbances of the solutions. Readings are made at wavelength 500 mμ, but wavelengths from 420 to 540 mμ give satisfactory readings with nesslerized solutions.

NOTE: Although photometric readings of the standard solutions show a reasonable constancy in repeated determinations when uniform conditions are provided, it is desirable to run concurrent standards with each group of unknowns.

Calculation

$$\frac{\text{Absorbance of unknown}}{\text{Absorbance of standard}} \times 30 = \text{milligrams of NPN per 100 ml. of blood}$$

Comments

In assessing the course of uremia, some workers have considered it to be more valuable to determine blood urea than to determine nonprotein nitrogen. Sometimes during the early stages of nephritis, variations in urea are more readily apparent than are changes in the NPN value. In actual practice, many physicians use blood urea or blood NPN determinations for the same clinical purpose, depending upon which determination is provided by the laboratory. A few conditions, such as eclampsia and hepatic failure, may show a disproportionate rise in blood NPN as compared with urea.

Range of Values

A little more than half of the nonprotein nitrogen of blood is derived from urea, uric acid, creatine and creatinine, ammonia, and amino acids. The remainder is made up of the nitrogen of glutathione and other peptides, purine and pyrimidine compounds, and unknow constituents. The normal values range from 25 to 40 mg./ 100 ml. of blood. In general, any condition resulting in a rise of blood urea will result in an increase in blood NPN. This is not only a result of rising blood urea but frequently is due also to retention of other constituents. Elevated levels as high as 400 mg./100 ml. have been reported in severe uremia.

Comparison of Results from Different Laboratories

In checking the NPN method for accuracy, the three cooperating laboratories analyzed a series of samples of blood filtrates and syn-

NONPROTEIN NITROGEN 105

TABLE I
NONPROTEIN NITROGEN VALUES OBTAINED BY THREE CHECKING LABORATORIES ON STANDARD SYNTHETIC SOLUTIONS*

Standard no.	Lab. A	Lab. B	Lab. C	Calculated
1	35	30	—	32
2	53	54	54	54
3	44	44	—	45
4	76	72	80	76
5	26	26	—	28
6	49	47	52	47
7	43	44	—	43
8	75	75	66	72
9	68	70	68	68
10	52	56	56	56

* These solutions were made up to contain known amounts of urea and creatinine with glucose added to simulate blood filtrates. Results are expressed in milligrams of nitrogen per 100 milliliters.

TABLE II
NONPROTEIN VALUES OBTAINED ON A SERIES OF BLOOD FILTRATES BY THREE CHECKING LABORATORIES*

Sample no.	Lab. A	Lab. B	Lab. C
1	31	33	35
2	31	31	32
3	57	64	65
4	41	48	48
5	80	102	97
6	33	34	34
7	38	40	43
8	60	66	61
9	44	51	53
10	47	50	49
11	36	36	35
12	37	37	37
13	101	105	114
14	33	35	37
15	—	60	60
16	116	122	120
17	231	232	217
18	49	47	48
19	230	265	230

* Results are expressed in milligrams of nitrogen per 100 milliliters.

thetic standards made up of urea, creatinine, and glucose in suitable concentration. These were shipped to each laboratory, and the results obtained by the three laboratories are presented in tables below. Table I contains the data obtained on standards of known nitrogen content, and Table II shows the results on blood filtrates.

It is apparent when comparing the work on standards that no one of the three laboratories appeared more skillful than the others in obtaining the theoretical amount of nitrogen. Agreement between the laboratories was generally good. Some divergence of results between the laboratories is attributed to different conditions and instrumentation or to the fundamental defects of the method. Our study shows the extent of such variability among three reliable stations. Comparison of the results on 19 blood filtrates, presented in Table II, further illustrates divergences between the different laboratories. As would be expected, the high-level NPN values usually show larger absolute differences, but these are not proportionately more excessive. It is clear that while the method is capable of yielding valuable clinical data when correctly applied, it does have very decided limitations in precision. Any failure to apply necessary precautions will obviously result in further loss of precision.

REFERENCES

1. Folin, O., and Wu, H., A system of blood analysis. *J. Biol. Chem.* **38**, 81–110 (1919).
2. Vanselow, A. P., Preparation of Nessler's reagent. *Ind. & Eng. Chem. Anal. Ed.* **12**, 516–517 (1940).
3. Hiller, A., and Van Slyke, D. D., A study of certain protein precipitants. *J. Biol. Chem.* **53**, 253–267 (1922).
4. Somogyi, M., Nitrogenous substances in zinc filtrates of human blood. *J. Biol. Chem.* **87**, 339–344 (1930).

DETERMINATION OF BLOOD pH

Submitted by: Jon V. Straumfjord, Jr., Clinical Biochemistry Laboratory, University Hospitals, State University of Iowa, Iowa City, Iowa

Introduction

The technique of measuring blood pH has improved in the past few years to a point where such measurements are being adopted as a routine clinical procedure. Definition of a standard procedure for pH determination is difficult because of the wide variations in the precision of commercially available apparatus and accessories. The optimal instrumentation for accurate measurement entails a cost ranging from $600 to $900. By lowering the cost to approximately $400 one compromises the accuracy and some of the potential effectiveness of the method, but even under these conditions severe abnormalities in blood pH can be detected. The basic principles of the measurement of pH will be discussed first, after which several proposed procedures will be described and evaluated.

The procedure for pH determination is simple yet fraught with subtle sources of significant error which easily escape attention. Clinical signs and symptoms to guide a physician in electrolyte aberrations are often limited; he is particularly dependent upon the laboratory for assistance in judging the proper therapy. An erroneous pH could well result in deleterious or even fatal intravenous fluid therapy. Because these subtle sources of error may occur in this potentially very important determination, it is necessary to exercise caution and constant evaluation of the accuracy. These sources of error will be discussed separately later.

The pH Scale

The formal relationship between pH and the hydrogen ion concentration as defined by Sørensen is $pH = -\log (H^+)$ where the log is to base 10 and (H^+) is the concentration of hydrogen ions. Actually the pH as measured is not a function of the hydrogen ion concentration, but rather is a function of the activity of the hydrogen ions. However, for most purposes this distinction is academic and is generally ignored.

The standard electrode for calibration of the pH scale is the hydrogen electrode. Stadie *et al.* (1) compared the pH of blood determined with the hydrogen electrode and with the glass electrode. They found that, even though the glass electrode was reproducible to 0.0016 pH units, there was approximately 0.01 pH unit difference between the glass and the hydrogen electrodes. Therefore, even though it is possible, with at least several pH meters, to read the pH to the third decimal place, there is doubt as to its significance. Certainly, under routine conditions, the blood pH measurement is not significant beyond the second decimal place. This should not imply that pH meters capable of reading to the third decimal place are not advantageous or necessary. Less sensitive meters are apt to decrease the significance to a greater extent.

The Glass Electrode and pH Meters

The glass electrode is prepared from a type of soft glass which, at pH values from 1 to 9, is essentially permeable only to hydrogen ions. The mechanism of this selective permeability is poorly understood, but it is linear, reproducible, and proportional to the pH of the solution being tested. The pH is determined by ascertaining the influence of hydrogen ions, transferred into the glass electrode, on the electromotive force (emf) between two half cells. The relationship of the glass electrode is shown by the word diagram Ag-AgCl | 0.1 N HCl | Glass | Blood | Saturated KCl | HgCl-Hg. The exact value of the emf between the half cells is not as important as its constancy because the measurement entails determining the difference in emf before and after the addition of hydrogen ions. This is analogous to colorimetric procedure where the difference is measured between a standard and an unknown. The change in emf is small, but after amplification it can be readily measured with a potentiometer previously standardized against a cell of known emf.

PH meters vary somewhat in their design and a description of all types is beyond the scope of this discussion. Specific information about various circuit types may be found in "Electrometric pH Determinations," by Bates (2). However, the general principles of operation of most meters suitable for blood pH are the same. They are composed of an amplifying circuit, a very sensitive potentiometer with a null-point indicator, and a pair of electrodes. Determination of pH necessitates several distinct operations with the meter. The precision of the final result is no better than the precision of each individual step. The manufacturer's instructions should be consulted for the method of operation of the individual pH meter.

Selection of the pH Meter and the Electrode

Several considerations influence the choice of equipment for blood pH determination. As was previously indicated, the expense is balanced against the accuracy desired. A wide variety of pH meters are available, but those recommended for blood pH are few. The meter should be accurate to 0.02 pH units. Sensitivity, reproducibility, and accuracy should be distinguished from one another. An instrument may be sensitive to a change of 0.005 pH units but be reproducible to only ±0.01 pH units within a series of determinations. The accuracy may be only 0.02 pH units because of slight errors in the linearity of the electrode and the calibration of the potentiometer. An over-all accuracy of ±0.02 to ±0.05 pH units can be used satisfactorily for routine clinical pH determinations. However, a limit of ±0.02 pH units instrumental error is needed to balance the other possible errors that may occur with handling of the blood, etc. The pH meters employing a potentiometer circuit with a null-point indicator are considered superior to the direct-reading type (2).

Several factors influence the choice of an electrode, such as the amount of blood required by the electrode, accuracy of the pH desired, and whether or not the electrode must fit into a special apparatus for constant-temperature regulation. The advantages of constant-temperature regulation will be discussed later.

Some of the various types of electrodes which can be used for blood pH are shown in Fig. 1. The shaded areas point out the similarities of the electrodes. Type 1 is the probe electrode commonly used in the laboratory. As generally employed, the probe electrode is inserted into blood in an open vessel. This procedure has the disadvantage of an alkaline error due to loss of CO_2 from the large exposed surface. This alkaline error is not as great in patients with a metabolic acidosis with a low carbon dioxide content of blood (3). The next two electrodes are especially designed for use with blood. Type 2 is the MacInnes-Belcher electrode which is used with the

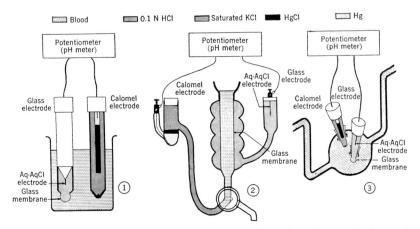

Fig. 1. A comparison of three types of glass electrodes. Type 1 is the probe electrode; type 2 is the MacInnes-Belcher electrode used with the Cambridge Model R pH meter; and type 3 is a Beckman Instrument Company electrode for use with the Model G or GS pH meters.

Cambridge Model R pH meter. The upper portion of the blood is exposed to air, but the surface area exposed is so small with respect to volume that no significant difference is noted in the pH whether the electrode is capped or left open. Type 3 may be secured from the Beckman Instrument Company for use with the Beckman Model G or GS meters. The Beckman electrodes have also been used satisfactorily with other pH meters (4).

Electrodes which may be connected directly to a syringe have been designed for determining the pH at the bedside. The blood is drawn directly into the electrode and the pH then determined. It must be kept in mind, however, that the warm blood may alter the temperature of the electrode. This creates problems of standardization and, unless corrected for, may induce errors which will be discussed later.

The pH of blood varies considerably with change in temperature because of its content of weak acids which show greater or lesser dissociation, depending upon the temperature of the environment. Rosenthal (5) has determined a correction factor that expresses the variation of the pH of human blood as a function of the temperature. Using his equation

$$\text{pH}_{37^\circ} = \text{pH} - (37^\circ - T)\, 0.0146$$

the pH can be calculated from that determined at any temperature between 18° and 38°C. However, this variation with temperature change is not the same in blood specimens from all individuals. The reason for variability in individual samples has not been clarified but the variation is likely due to a combination of factors such as differences in ion concentration and in number of cells in blood. Craig *et al.* (6) noted that the temperature factor varied from 0.0103 to 0.0193 in different individuals but reported a mean factor of 0.0149 ± 0.00273, which is in agreement with Rosenthal's factor. The error induced by using one standard deviation from Craig's mean factor is approximately ±0.03 pH units, and Craig's maximum limits indicate a potential error of ±0.05 pH units. The temperature-correction dial which is present on most pH meters does not compensate for the effect of temperature on the pH of blood. To obtain the highest accuracy of blood pH determinations, an electrode maintained at 37.5°C. is required.

The electrode maintained at room temperature can be used successfully in a routine clinical laboratory when the temperature correction is made, but a loss of potential effectiveness must be accepted. If all the potential errors of a room-temperature electrode should be additive, the total error would approach ±0.09 pH units. This is an unlikely situation. The true limits of error are difficult to predict because of the variations in equipment and individuals, but are approximately ±0.05 pH units. The limits of error with a constant-temperature electrode at 37.5°C. are approximately ±0.025 pH units or less, depending upon the equipment used.

Incubators (7, 8), water baths (9, 10), or an oil bath in a heated compartment (11) have been used for maintaining an electrode at body temperature. The author has found a water-jacketed electrode (produced by the Cambridge Instrument Company) with water at 37.5°C. circulating around the electrode to be very satisfactory. Figure 2 illustrates the type of apparatus used in our laboratory. Only the glass electrode is maintained at 37.5°C. with this method. Although the calomel electrode and the salt bridge are maintained at room temperature, they should be protected from drafts and other sources of sharp temperature change. The apparatus for constant-temperature regulation varies in cost from approximately $50 to $450, depending upon the type used and ingenuity of the individual assembling it. The temperature regulation should be within 0.1°C.

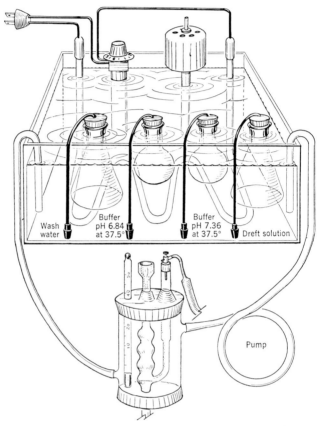

Fig. 2. Apparatus for the maintenance of the glass electrode, buffers, wash water, and cleaning solution at 37.5°C.

Care of the Electrode

The electrode should be kept clean. The routine rinse, with a neutral detergent such as Dreft (1% solution), following the use of blood is helpful in preventing excessive contamination of the electrode. Mineral oil should be kept from the electrode. When not in use, the electrode should be kept in contact with distilled water. For prolonged storage 0.1 N HCl is recommended (2).

Colorimetric Determination of Blood pH

Several methods are available for the colorimetric determination of blood pH (12). The reported accuracy with these procedures is satisfactory for routine clinical use. They are, nevertheless, much more time-consuming than the electrometric method and, because there is more manipulation of the blood, there is potentially a greater source of error.

Collection and Preparation of the Blood

A simple method for the collection of blood follows: A sterile 5- or 10-ml. syringe with a tight-fitting needle and plunger is coated with heparin by aspirating approximately 0.5 ml. of heparin solution (10 mg./ml.) and manipulating the plunger so that most of the surface of the interior is coated. The excess is then expelled. This gives sufficient protection against air contamination of the blood. No significant difference has been found in the pH of blood drawn in syringes prepared in this manner as compared with blood drawn in syringes prepared with heparin and a thin coat of mineral oil. The amount of blood required depends upon the type of electrode used and varies from several drops to 3 ml.

A tourniquet may be used while drawing venous blood, but all equipment should be prepared in advance so there is no unnecessary delay in obtaining blood after stasis is applied. The effect of the tourniquet was studied in five normal volunteers with veins suitable for venepuncture without a tourniquet. There was no difference between the control pH and the pH of the sample removed after the tourniquet had been in place for 2 minutes if the blood was drawn with the tourniquet still in place. Stasis induced by the tourniquet alters the pH of the blood at the capillary level. Since there is little metabolic activity within the lumen of the antecubital vein, the pH of the blood collected from this site while the tourniquet is in place is only minimally affected by metabolic end products. For this reason, blood should be drawn with the tourniquet in place.

Hemolysis is to be avoided. The liberation of carbonic anhydrase from the red cells may alter the pH by favoring destruction of H_2CO_3 and $NaHCO_3$ (3).

The blood should be drawn anaerobically and with minimal negative pressure. However, if a slight amount of air contamination does occur, no significant alteration is noted if the air is promptly expelled. A small cork placed over the needle prevents spillage and air exposure of the blood en route to the laboratory. Air contamination of blood results in loss of carbon dioxide, and the pH becomes more alkaline. Figure 3A demonstrates the change in pH when blood is

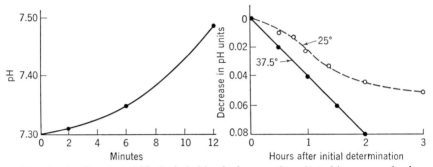

Fig. 3. *A.* Change in pH of whole blood when gently agitated in an open beaker at 25°C. *B.* Decrease in the pH of whole blood with time when incubated anaerobically at 25° and 37.5°C. All pH determinations were made at 37.5°C.

gently agitated in a small, uncovered beaker. Shed blood maintained anaerobically becomes progressively more acid owing predominantly to the production of lactic acid by erythrocytes and leucocytes. Figure 3B shows the drop in pH with respect to time. Determination of the pH within 15 minutes after the blood has been drawn is therefore advisable. If the blood pH cannot be determined immediately, the production of lactic acid may be curbed in several ways. A special pipet has been designed so that blood may be safely collected with 1% sodium fluoride to inhibit lactic acid production (6). The syringe may be stored at 4°C., a temperature which largely, but not completely, inhibits glycolysis. Wilson (7) reported a decrease of less than 0.03 pH units when whole blood was stored at 4°C. for 5 hours. However, the blood stored at this temperature must be warmed to the temperature of the electrode prior to pH determination.

Unlike whole blood, plasma stored anaerobically shows no decrease in pH with time. However, the pH of plasma is affected by the temperature at which it is separated from the cells. Consequently, to gain a true reflection of whole-blood pH, the plasma must be separated from the cells at 37.5°C. (5). This error is not corrected by measuring the pH of the plasma at 37.5°C.

Reagents

1. *Disodium hydrogen phosphate, anhydrous* C.P. (Na_2HPO_4). This compound is somewhat hygroscopic and must be dried for 2 hours at 120°–130°C. prior to use (2).

2. Potassium dihydrogen phosphate, C.P. (KH_2PO_4).
3. Potassium hydrogen phthalate, C.P. ($KHC_8H_4O_4$).
4. Sodium borate, C.P. ($Na_2B_4O_7.10H_2O$).

PREPARATION OF STANDARD BUFFERS (2)

A standard buffer, pH 7.38 at 25°C., with an ionic strength of
0.1 is prepared by adding 1.246 g. of potassium dihydrogen phosphate and 4.30 g. disodium hydrogen phosphate to 1 l. of freshly
boiled distilled water at 25°C. This buffer has a pH of 7.36 at 37.5°C.

A standard buffer pH 6.86 at 25°C. with an ionic strength of 0.1
is prepared by the addition of 3.40 g. of potassium dihydrogen phosphate and 3.55 g. of disodium hydrogen phosphate to 1 l. of freshly
boiled distilled water at 25°C. This buffer has a pH of 6.84 at 37.5°C.

The addition of 10.1 g. of potassium hydrogen phthalate to 1 l.
of freshly boiled distilled water will give a 0.05 M standard buffer
of pH 4.01 at 25°C. and pH 4.025 at 37.5°C.

The addition of 3.81 g. of sodium borate to 1 l. of freshly boiled
distilled water at 25°C. will give a 0.01 M standard buffer of pH
9.18 at 25°C. and pH 9.086 at 37.5°C.

Bates (2) recommends buffers 2 pH units apart for checking the
linearity of the glass electrode. For this reason the phthalate and
borax buffers are included. The phosphate buffers are for routine
use. Only periodic checks with the phthalate and borax buffers are
necessary.

It is recommended that fresh buffers be prepared every 2–3
months. Sterilization of the containers is not necessary. These buffers
have stood for 17 months with as little as 0.007 unit change in
pH (2). These chemicals may be secured from the National Bureau
of Standards.

Procedure

GENERAL METHOD

1. Turn on the meter and warm it up prior to drawing the blood.
2. Draw blood anaerobically with a syringe prepared with heparin,
as previously outlined. The amount of blood needed depends upon the
electrode used. A tourniquet may be used during the collection of
venous blood. However, the blood should be drawn with the tourniquet in place.
3. The buffers, wash water, and cleaning solution should be within

1°C. of the temperature of the electrode. The use of solutions at temperatures widely different from that of the electrode will introduce significant error. Keep the saturated potassium chloride at room temperature if the salt bridge and calomel electrode are at room temperature.

4. Balance the meter according to the manufacturer's instructions. Then standardize it with a standard buffer.

5. Rinse the electrode well with distilled water. Determine the responsiveness of the electrode by measuring the pH of another standard buffer at least 0.4 pH units different from the first. The standard buffers should check within ±0.01 pH units. Wash the electrode again with distilled water.

6. If a room-temperature electrode is used, allow the blood to cool to room temperature prior to placing the blood in the electrode. After the blood has been placed in the electrode, the meter may initially drift slightly. This drift is followed by stabilization in approximately 45–60 seconds. The stabilized reading is considered the true pH. Differences in temperature of the blood and electrode will induce drift, but a slight amount of initial drift will occur even if both electrode and blood are at the same temperature. The nature of this drift is uncertain. Then determine the pH by balancing the instrument to the null point. Three separate readings should be taken and averaged. If the null point should drift, see discussion of sources of error. The pH is corrected to 37.5°C. with the equation

$$pH_{37.5} = pH_t - (37.5 - t)\, 0.0146$$

where t = electrode temperature.

7. If an electrode at 37.5°C. is used, warm the blood to 37.5°C. prior to placing it in the electrode. After placing it in the electrode, allow from 45 seconds to 1 minute to elapse before determining the pH. Take three separate readings. Report the average pH.

MICROMETHOD WITH OPEN ELECTRODE (DRYER (13))

1. Turn on the pH meter and allow it to warm up. Standardize the meter with standard buffers.

2. Carefully rinse off the electrodes with a little distilled water to remove traces of buffer. Wipe with cleaning tissue or a soft cloth.

3. Half fill a sample cup with physiological saline and measure

the pH, which will usually be found to be in the neighborhood of 6.4–6.8.

4. By allowing traces of the fumes either of acid (usually HCl) or alkali (usually NH₃) to enter the saline, adjust the pH of the saline to between 7.2 and 7.4. Values outside this range are not permissible. This procedure is best accomplished by waving the bottle stopper, wet with acid or alkali, over the top of the cup. Be sure to allow good mixing of the vapor with the saline in the cup.

5. Add 3 or 4 drops of blood from the syringe to the adjusted saline. Mix quickly by swirling the beaker and promptly place the beaker in position in the instrument. After 45–60 seconds of equilibration, measure the pH.

6. Correct the measured pH by noting the electrode temperature of the measurement and applying the temperature correction formula

$$pH_{37.5} = pH_{t°} - 0.0146 \ (37.5 - t°).$$

The normal range for venous blood with this method is 7.35 to 7.45.

Detection of the Source of Error

Sources of error can be subdivided into two groups. One involves malfunction of the instrument; the other, the handling of blood.

The routine use of two standard buffers of different pH values will indicate malfunction of the equipment and does more to indicate or prevent error than any other procedure. If the measured pH's of the buffers agree within ±0.01 units, then the chief sources of error are in the manipulation of the blood or in an improper connection in the blood column or salt bridge. If the pH's of the buffers do not agree, then the fault is generally attributable to one of the following sources (14):

1. Dirty electrode.
2. Cracked electrode.
3. Faulty standard cell in the pH meter.
4. Cracked or broken assembly of the glass of the calomel electrode.
5. Contamination of the potassium chloride solution from the calomel electrode or salt linkage.
6. Interruption of the flow of potassium chloride solution from the calomel electrode.

7. Inaccurate buffer solutions.

8. Worn-out batteries.

9. Imperfect or incorrect battery connections.

10. Defective or worn-out tubes.

Generally if the fault lies in the meter, it will be impossible or difficult to balance it. The manufacturer's instructions should be consulted for locating difficulties within the meter.

In our experience the two most common sources of error have been (a) drifting due to contamination and moisture around the electrode connections and (b) an improper connection in the salt bridge between the blood and the calomel electrode. Occasionally the null-point indicator will drift slightly after it has been set against the standard buffer. If this occurs, the meter should be restandardized. If a slight drift is observed after the blood has been placed in the electrode, less error will result if the pH dial is balanced to the drifted point rather than to the original point on the null-point indicator. The use of buffers or wash water at a temperature widely different from that of the electrode will cause a significant error (8, 2). This may be noticed when solutions at room temperature are used in an electrode maintained at 37.5°C. Therefore, careful standardization of the instrument, careful routine inspection of the blood-salt bridge connections, and the routine use of buffers of two pH values will prevent the great majority of instrumentation errors that can occur with blood pH determination.

Sources of error not directly related to instrumentation are air contamination of the blood, extensive hemolysis, use of too much anticoagulant, and failure to use the temperature correction factor when indicated. As was previously pointed out, considerable error may occur in the pH if blood is allowed to stand for appreciable time. Unfortunately, there is no readily available laboratory method to determine the extent of this error. Therefore, if the freshness of the sample is in doubt, the determination should be rerun with fresh blood.

Severinghause et al. (11) have reported that small errors may occur owing to slight deviation in the linearity of the glass electrode and the potentiometer. If extremely accurate pH determinations are needed, the electrode should be calibrated with 3 buffers at pH values of approximately 4, 7, and 9. The potentiometer may be calibrated with an accurate source of known voltage derived from a standard potentiometer.

Range of Values

The reported normal arterial and venous pH range depends upon the technique employed. The more rigidly controlled methods will define narrower limits for the normal. In our laboratory, in which a constant-temperature electrode maintained at 37.5°C. is used with a very sensitive meter, the normal venous pH is 7.34 ± 0.025 pH units, with 7.27 and 7.39 as lower and upper limits in 32 individuals with a total of 44 determinations. Astrup and Schroder (10) have reported the normal pH of venous blood in 23 men and women to vary from 7.28 to 7.35, with a mean of 7.32 ± 0.018. Our series of normal arterial pH values have a mean of 7.42 ± 0.012. This series agrees well with that of Wilson (7) who reported a normal mean of 7.40 ± 0.003.

Clinical Interpretation

It is not the purpose of this section to discuss all the changes in the hydrogen ion concentration in disease. This is adequately discussed elsewhere (15, 16). It should be pointed out, however, that the measurement of pH determines only the direction and degree of alteration of the hydrogen ion concentration and not the cause. Other determinations are necessary along with clinical findings to complete the acid-base story. The blood pH in conjunction with the CO_2 content will indicate the degree of an individual's compensation to an acid-base aberration and will help indicate the presence of mixed alterations in electrolyte status. Completely erroneous conclusions in mixed alterations occur when only total CO_2 content is determined. The pH measurement is especially useful in these cases. Construction of a chart indicating the relationship between total CO_2 and pH (16) is helpful in elucidating the extent of a patient's compensation.

The relationship between P_{CO_2} and total CO_2 serves as a check on the pH measurement. By using the Henderson-Hasselbalch equation modified for P_{CO_2}

$$pH = 6.10 + \log_{10} \frac{\text{total } CO_2 - 0.0301 \, P_{CO_2}}{0.0301 \, P_{CO_2}} \qquad (16),$$

one can calculate the pH. The calculated pH and the determined pH show very good agreement, especially if corrections are made for the changes in the pK' as pointed out by Severinghaus *et al.* (17). The pH determination can serve as an extremely valuable adjunct to the laboratory studies and is easily adapted to routine clinical use. In many cases in which the acid-base status is not clarified by total CO_2, Na, K, and Cl, the pH will do more than any other procedure to make sense out of chaos.

REFERENCES

1. Stadie, W. C., O'Brien, H., Laug, E. P., Determination of the pH of serum at 38° with the glass electrode and an improved electron tube potentiometer. *J. Biol. Chem.* **91**, 243–269 (1931).
2. Bates, R. G., "Electrometric pH Determinations," pp. 262, 118–121, 290, 251, 284. Wiley, New York, 1954.
3. Natelson, S., and Tietz, N., Blood pH measurement with the glass electrode. *Clin. Chem.* **2**, 320–327 (1956).
4. Seligson, D., personal communication (1956).
5. Rosenthal, T. B., The effect of temperature on the pH of blood and plasma in vitro. *J. Biol. Chem.* **173**, 25–30 (1948).
6. Craig, F. A., Lange, K., Oberman, J., and Carson, S., A simple, accurate method of blood pH determinations for clinical use. *Arch. Biochem. Biophys.* **38**, 357–364 (1952).
7. Wilson, R. H., pH of whole arterial blood. *J. Lab. Clin. Med.* **37**, 129–132 (1951).
8. Holaday, D., An improved method for multiple rapid determinations of arterial blood pH. *J. Lab. Clin. Med.* **44**, 149–159 (1954).
9. Murray, J. T., Determination of blood pH and a new cell for blood pH. *Am. J. Clin. Pathol.* **26**, 83–89 (1956).
10. Astrup, P., and Schroder, S., Apparatus for anaerobic determination of the pH of blood at 38 degrees centigrade. *Scand. J. Clin. & Lab. Invest.* **8**, 30–32 (1956).
11. Severinghaus, J. W., Stupfel, M., and Bradley, A. F., Accuracy of blood pH and Pco_2 determinations. *J. Appl. Physiol.* **9**, 189–196 (1956).
12. Van Slyke, D. D., Weisiger, J. R., and Keller, K., Photometric measurements of plasma pH. *J. Biol. Chem.* **179**, 743–756 (1949).
13. Dryer, R. L., unpublished method.
14. Instructions for servicing Beckman Model G pH meter. Beckman Bulletin 132-D. Beckman Instruments, Inc., Fullerton, Calif.
15. Gamble, J. L., "Companionship of Water and Electrolytes in the Organization of Body Fluids." Stanford Univ. Press, Stanford, California, 1951; Hardy, J. D., "Fluid Therapy." Lea and Febiger, Philadelphia, Pennsylvania, 1954; West, E. S., and Todd, W. R., Chemistry of respiration, acid base balance, and electrolyte and water balance. "Textbook of Biochemistry," pp. 585–654. Macmillan, New York, 1955.

16. Davenport, H. W., "The ABC of Acid-Base Chemistry," 3rd ed., pp. 49, 29. Univ. of Chicago Press, Chicago, Illinois, 1950.

17. Severinghaus, J. W., Stupfel, M. and Bradley, A. F., Variations of serum carbonic acid pK' with pH and temperature. *J. Appl. Physiol.* **9,** 197–200 (1956).

Additional references:

Consolazio, C. F., Johnson, R. E., and Marek, E., Hydrogen ion concentration. "Metabolic Methods—Clinical Procedure in the Study of Metabolic Function," pp. 234–243. Mosby, St. Louis, Missouri, 1951.

Dole, M., "The Glass Electrode. Methods, Applications, and Theory." Wiley, New York, 1941.

Koch, F. C., and Hanke, M. E., Hydrogen ion activity and pH. "Practical Methods in Biochemistry," 6th ed., pp. 98–132. Williams & Wilkins, Baltimore, Maryland, 1953.

Singer, R. B., A new diagram for the visualization and interpretation of acid-base changes. *Am. J. Med. Sci.* **221,** 199–210 (1951).

ALKALINE AND ACID PHOSPHATASE*

Submitted by: MARGARET M. KASER, Veterans Administration Center, Wood, and Marquette University School of Medicine, Milwaukee, Wisconsin

JOHN BAKER, Veterans Administration Center, Wood, Wisconsin

Checked by: BERNARD LONGWELL, Department of Biochemistry, Lovelace Clinic, Albuqerque, New Mexico

MIRIAM REINER and HELEN L. CHEUNG, District of Columbia General Hospital, Washington, D. C.

Introduction

Although the presence in mammalian blood and tissues of enzymes capable of hydrolyzing glycerophosphate had been demonstrated previously by several investigators, Kay (1, 2) was the first to describe a method for the quantitative determination of plasma phosphatase activity using this substrate. Sodium β-glycerophosphate and unbuffered oxalated plasma were incubated for 48 hours at pH 7.6 and 38°C., and the phosphorus in a trichloroacetic acid filtrate was determined by the method of Briggs (3) or by some other procedure. Bell and Doisy (4) had devised a method for the estimation of phosphorus in blood which did not require the preliminary isolation of the phosphorus as ammonium phosphomolybdate but depended upon the production of a blue color with phosphate in the presence of an excess of molybdate ions and with hydroquinone as a reducing agent. Briggs (3) improved the quality and the stability of the color by the addition of sodium sulfite and by carrying out the reaction in an acid medium. Bodansky (5) modified Kay's procedure by the use of a glycerophosphate substrate in Veronal buffer at pH 8.6 and by reducing the period of incubation to 1 hour. The phosphorus liberated was determined by the Kuttner-Cohen-Lichtenstein procedure as modified by Raymond and Levene (6–10), in which stannous chloride was substituted for hydroquinone to give

* Based on the methods of Bodansky (5), Shinowara, Jones, and Reinhart (11) and Fiske and SubbaRow (12).

122

greater sensitivity and much more rapid color development. However, Bodansky found corrections necessary for the deviation of the
color from Beer's law and for the effect of the trichloroacetic acid
and glycerophosphate in the filtrate. He defined the unit of alkaline
phosphatase activity as the degree of activity represented by the
liberation of 1 mg. of inorganic phosphorus by 100 ml. of serum from
the sodium β-glycerophosphate substrate at pH 8.6 and 37°C. during
the first hour.

Shinowara, Jones, and Reinhart (11) investigated the Bodansky
procedure for alkaline phosphatase quite carefully, especially with
respect to the optimum pH for enzyme activity, which they found
to be pH 9.3 instead of 8.6 at 37°C. The procedure presented here
includes a modification in the preparation of substrate to give a final
pH of 9.3. However, the phosphorus is determined by the method
of Fiske and SubbaRow (12), in which 1-amino-2-naphthol-4-sul-
fonic acid with sodium sulfite and sodium bisulfite is employed as
the reducing reagent rather than the stannous chloride of the Kuttner-Cohen-Lichtenstein procedure used by Shinowara, Jones, and
Reinhart. Furthermore, the quantities of the samples and reagents
have been altered to yield less intensely colored solutions of suitable
absorbance for photometry. Shinowara, Jones, and Reinhart (11)
also adapted the Bodansky procedure to the determination of serum
acid phosphatase, which has a pH optimum of 5.0. According to
their data the pH optima of the two types of enzyme are distinct,
and there is practically no overlapping of their activity curves. The
unit of acid phosphatase activity is similar to that for alkaline phosphatase and is equal to 1 mg. of inorganic phosphorus liberated from
the sodium β-glycerophosphate substrate at 37°C. and pH 5.0 in
one hour by 100 ml. of serum.

Phosphorus may be determined by the method of Power and
Young described in Volume I of this series (13) with Elon (p-methylaminophenol) as the reducing agent. One of the checkers (M. R.)
has compared this procedure with that of Fiske and SubbaRow
and has obtained identical results.

Reagents

1. Trichloroacetic acid, 30%. Dissolve 300 g. of reagent grade trichloroacetic acid in water and dilute to 1 l.

2. Acid molybdate reagent. This reagent consists of 2.5% ammonium molybdate in 3 N sulfuric acid. Dissolve 25 g. of reagent grade ammonium molybdate in approximately 500 ml. of water in a 1-l. volumetric flask, add 300 ml. of 10 N sulfuric acid, and dilute to volume with water. To prepare the 10 N sulfuric acid, add 278 ml. of concentrated sulfuric acid slowly with mixing to about 700 ml. of water in a Pyrex beaker or flask. Cool to room temperature and dilute to 1 l. with water.

3. *Aminonaphtholsulfonic acid, 0.25%.* The 1-amino-2-naphthol-4-sulfonic acid used for the preparation of this reagent is grayish or light pink in color. If there is too much color, it is desirable to remove the excess color. This can often be done satisfactorily by placing the material on filter paper in a Buchner funnel and washing it with small amounts of ethanol. If necessary, the compound can be recrystallized as follows: Add 15 g. of the crude sulfonic acid to a liter of water at 90°C., which contains 150 g. of sodium bisulfite and 10 g. of crystalline sodium sulfite ($Na_2SO_3.7\ H_2O$). Filter the solution when all but the amorphous solid has dissolved. Cool the filtrate and add to it 10 ml. of concentrated hydrochloric acid. Filter with suction and wash the precipitate with about 300 ml. of water and then with alcohol until the washings are colorless. The sulfonic acid should be air-dried with as little exposure to light as possible and then kept in a dark bottle.

Dissolve 0.50 g. of 1-amino-2-naphthol-4-sulfonic acid in 195 ml. of 15% sodium bisulfite to which 5 ml. of 20% sodium sulfite is added. More sodium sulfite may be needed to effect solution. If so, it should be added 1 ml. at a time followed by shaking. However, no more sodium sulfite, which increases pH, should be added than is necessary since the reagent is more stable in solutions of high acidity. The solution is stable for several weeks if kept in the refrigerator and protected from exposure to air and light.

4. *Stock substrate.* Dissolve 5.0 g. of reagent grade sodium β-glycerophosphate and 4.24 g. of sodium barbital in water and dilute to 500 ml. This reagent, if covered with a layer of petroleum ether in a glass-stoppered bottle, will keep in the refrigerator for one month.

5. *Alkaline working substrate.* Use for the alkaline phosphatase determination. To 100 ml. of the stock substrate add 1.4–2.0 ml. of 0.1 N sodium hydroxide and dilute with water to 200 ml. The pH of this solution should be 9.70–9.75 at 25°C.

6. Acid working substrate. Use for the acid phosphatase determination. To 100 ml. of the stock substrate add 10.0 ml. of 1.0 N acetic acid and dilute with water to 200 ml. The pH of this substrate should be 5.00 ± 0.05.

7. Stock phosphorus standard, 1 mg. P/ml. Dissolve 0.4394 g. of potassium dihydrogen phosphate (KH_2PO_4) in 100 ml. of water.

8. Dilute phosphorus standard, 0.01 mg. P/ml. Dilute 1 ml. of the stock phosphorus standard to 100 ml. with water.

Procedures

STANDARDIZATION

Prepare a series of test tubes as indicated in Table I. To each tube add 1.0 ml. of 30% trichloroacetic acid, 1.0 ml. of acid molybdate reagent and 0.40 ml. of aminonaphtholsulfonic acid reagent. Mix well and allow to stand for 10 minutes to obtain full color development. With the blank tube in the series shown in Table I, adjust the galvanometer of the photometer to 100% transmittance at 660 mμ or with a red filter. Read the transmittance or absorbance values of the colored solutions. Construct a calibration curve by plotting the readings obtained against the corresponding values for serum inorganic phosphorus on semilog or coordinate paper.

If preferred, the color may be developed in 10-ml. glass-stoppered

TABLE I

QUANTITIES USED IN TEST TUBES PREPARED FOR
STANDARDIZATION PROCEDURE

Dilute phosphorus standard (ml.)	Phosphorus (mg.)	Water (ml.)	Equivalent serum inorganic phosphorus (mg./100 ml.)
0.50	0.005	7.50	2.0
1.00	0.010	7.00	4.0
1.50	0.015	6.50	6.0
2.00	0.020	6.00	8.0
2.50	0.025	5.50	10.0
3.00	0.030	5.00	12.0
3.50	0.035	4.50	14.0
4.00	0.040	4.00	16.0
0	0	8.00	0.0

volumetric flasks instead of in photometer cuvettes. Should this be done, the flasks should be partly filled with water before the addition of the reagents. The colored solutions are then diluted to a final volume of 10.0 ml. instead of 10.4 ml. In either event, follow the same procedure for the calibration curve as is used for samples. Identical curves are obtained whether the phosphate standards are made up in water or in appropriate concentrations of either the alkaline or the acid working substrate.

PROCEDURE FOR ALKALINE PHOSPHATASE

1. Serum inorganic phosphorus. To 7.50 ml. of alkaline working substrate in a test tube add 0.50 ml. of serum and 2.0 ml. of 30% trichloroacetic acid. Mix well and filter through a retentive, acid-washed filter paper. Transfer 5.0 ml. of filtrate to a photometer cuvette.

2. Total inorganic phosphorus. Warm 7.75 ml. of alkaline working substrate by immersing the tube in a water bath at 37°C. for 10 minutes. Add 0.25 ml. of serum, mix, and continue to incubate the tube at 37° for one hour. Add 2.0 ml. of 30% trichloroacetic acid, mix well, and filter. Transfer 5.0 ml. of filtrate to a photometer cuvette.

3. Reagent blank. Into a photometer cuvette measure 4.0 ml. of working substrate and 1.0 ml. of 30% trichloroacetic acid.

To all three cuvettes add 1.0 ml. of acid molybdate reagent, 0.4 ml. of aminonaphtholsulfonic acid reagent, and 4.0 ml. of water and mix. After 10 minutes adjust the galvanometer with the reagent blank to read 100% transmittance at 660 mμ, or with a red filter, and read the transmittance or absorbance values of the colored solutions. If desired, the color may be developed in 10-ml. glass-stoppered volumetric flasks as described in the procedure for standardization.

Determine the amount of inorganic phosphate present in the two solutions as serum inorganic phosphorus by reference to the calibration curve prepared in the standardization procedure and multiply the figure found for the incubated sample by 2 to correct for the smaller serum sample used. The alkaline phosphatase activity in Bodansky units is equal to the difference in inorganic phosphate values for the incubated and the unincubated samples.

PROCEDURE FOR ACID PHOSPHATASE

1. Serum inorganic phosphorus. Measure into a test tube 7.50 ml. of acid working substrate, 0.50 ml. of serum, and 2.0 ml. of 30% trichloroacetic acid. Mix well and filter through a retentive, acid-washed filter paper. Transfer 5.0 ml. of filtrate to a photometer cuvette.

2. Total inorganic phosphorus. Place 7.50 ml. of acid working substrate in a test tube and immerse the tube in a water bath at 37°C. for 10 minutes. Add 0.50 ml. of serum, mix well, and allow the tube to remain in the water bath for one hour. Then add 2.0 ml. of 30% trichloroacetic acid, mix well, and filter. Place 5.0 ml. of the filtrate in a photometer cuvette.

3. Reagent blank. Into a photometer cuvette measure 4.0 ml. of acid working substrate and 1.0 ml. of 30% trichloroacetic acid.

To each cuvette add 1.0 ml. of acid molybdate reagent, 0.40 ml. of aminonaphtholsulfonic acid reagent, and 4.0 ml. of water. Mix the solutions and after 10 minutes adjust the galvanometer of the photometer with the reagent blank to 100% transmittance at 660 mμ, or with a red filter, and determine the transmittance or absorbance of the colored solutions. As in the case of alkaline phosphatase, the color may be developed in 10-ml. glass-stoppered volumetric flasks, if desired.

Determine the inorganic phosphate in the incubated and unincubated samples by reference to the phosphorus calibration curve. The acid phosphatase activity is equal to the difference between the two results.

Discussion

In the early development of phosphate chemistry, phosphate was measured as phosphate and reported as phosphorus or phosphorus pentoxide. This tradition of measuring phosphate and reporting phosphorus has been carried on through the years. While this practice is not necessarily a good one, it has been followed in this report.

Blood for phosphatase determinations should be taken from fasting subjects. This is especially important for alkaline phosphatase. For alkaline phosphatase the serum may be preserved by freezing, but acid phosphatase is much less stable, and no attempt should be made to preserve it for any length of time. Acid phosphatase is readily inactivated at body or room temperature (14). If determinations cannot be carried out immediately after collection of the samples, the serum may be kept in an ice bath for an hour or two. Hemolysis of blood samples is to be avoided, since this will increase the phosphate content of serum because the red blood cells contain an acid phosphatase. A specific test for acid phosphatase is mentioned below.

All glassware used for the preparation and storage of reagents and for phosphatase determinations must be carefully cleaned. The method used for the determination of phosphate is very sensitive, and contamination of glassware with phosphate may lead to erroneous results. Some cleansing agents recommended for laboratory glassware contain trisodium phosphate or other phosphates. Thorough rinsing with distilled water must be assured if they are used.

In order for the amount of inorganic phosphate liberated from the substrate to be linearly related to the concentration of enzyme present, there must be a large excess of substrate, so that at the completion of the incubation period only a small fraction of the total amount of substrate shall have been hydrolyzed. If an alkaline phosphatase value above 20–30 units is encountered, the serum should be diluted with 0.9% sodium chloride solution and the entire test should be repeated. It is not sufficient to dilute the protein-free filtrate nor the colored solution. Of course, if the serum is diluted, the appropriate correction must be made in calculating the amount of inorganic phosphorus in the protein-free filtrate from the incubated sample.

In addition to glycerophosphate a number of other substrates have been proposed for the determination of phosphatase activity. The method of King and Armstrong (15, 16) has been widely used for acid phosphatase. Serum is incubated with phenyl phosphate, and the phenol liberated from the substrate is determined. Other substrates which have been suggested include p-nitrophenyl phosphate (17, 18), phenolphthalein phosphate (19), and sodium β-naphthyl phosphate (20). An advantage of the procedure described here is that an estimation of the serum inorganic phosphate is obtained as well as a measure of phosphatase activity.

Range of Values

Values for serum alkaline phosphatase for healthy adults fall between 1.0 and 6.0 units per 100 ml. of serum, but children may have activities as high as 15 units with a gradual decrease until puberty. The determination of alkaline phosphatase is useful in the diagnosis of abnormalities of parathyroid function and in a number of diseases of bone. However, for the proper interpretation of alkaline phosphatase values, other data relating to calcium and phosphorus metabolism must also be considered. These subjects have been well reviewed by Bodansky (21) and Hoffman (22). Serum alkaline phosphatase may be markedly elevated in some diseases of the liver, especially in obstructive jaundice (23, 24), and it may be increased in neoplastic involvement of the liver (23, 25, 26).

Because the enzymatic reaction is carried out at the optimum pH of 9.3, the amount of hydrolysis of the glycerophosphate is greater than at the pH of 8.6 recommended by Bodansky. The units of alkaline phosphatase as defined by Bodansky, but obtained at this higher pH, are approximately one-third greater than those obtained at pH 8.6.

The only well-established clinical indication for the determination of serum acid phosphatase is carcinoma of the prostate. In the presence of metastases a considerable proportion of patients will exhibit elevation of serum acid phosphatase (27, 28, 29, 30). Abul-Fadl and King (31) showed that "prostatic" acid phosphatase differed from other acid phosphatases in its sensitivity to tartrate. Fishman et al. (32, 33, 34) have published data which indicate that tartrate inhibition of prostatic phosphatase may be used to greatly improve the diagnostic value and specificity of the test. The so-called prostatic acid phosphatase determined by tartrate inhibition may be a significant contribution in this field. Tartrate can be added to the substrate above (34).

REFERENCES

1. Kay, H. D., Plasma phosphatase. I. Method of determination. Some properties of the enzyme. *J. Biol. Chem.* **89**, 235–247 (1930).
2. Kay, H. D., Plasma phosphatase. II. The enzyme in disease, particularly in bone disease. *J. Biol. Chem.* **89**, 249–266 (1930).
3. Briggs, A. P., A modification of the Bell-Doisy phosphate method. *J. Biol. Chem.* **53**, 13–16 (1922).
4. Bell, R. D., and Doisy, E. A., Rapid colorimetric methods for the determination of phosphorus in urine and blood. *J. Biol. Chem.* **44**, 55–67 (1920).
5. Bodansky, A., Phosphatase studies. II. Determination of serum phosphatase. Factors influencing the accuracy of the determination. *J. Biol. Chem.* **101**, 93–104 (1933).

6. Bodansky, A., Phosphatase studies. I. Determination of inorganic phosphate. Beer's Law and interfering substances in the Kuttner-Lichtenstein method. *J. Biol. Chem.* **99,** 197–206 (1932–1933).

7. Bodansky, A., Determination of serum inorganic phosphate and of serum phosphatase. *Am. J. Clin Pathol., Tech. Suppl.* **1,** 51–59 (1937).

8. Kuttner, T., and Cohen, H. R., Micro colorimetric studies. I. A molybdic acid, stannous chloride reagent. The micro estimation of phosphate and calcium in pus, plasma, and spinal fluid. *J. Biol. Chem.* **75,** 517–531 (1927).

9. Kuttner, T., and Lichtenstein, L., Micro colorimetric studies. II. Estimation of phosphorus: Molybdic acid-stannous chloride reagent. *J. Biol. Chem.* **86,** 671–676 (1930).

10. Raymond, A. L., and Levene, P. A., Hexose phosphates and alcoholic fermentation. *J. Biol. Chem.* **79,** 621–635 (1928).

11. Shinowara, G. Y., Jones, L. M., and Reinhart, H. L., The estimation of serum inorganic phosphate and "acid" and "alkaline" phosphatase activity. *J. Biol. Chem.* **142,** 921–933 (1942).

12. Fiske, C. H., and SubbaRow, Y., The colorimetric determination of phosphorus. *J. Biol. Chem.* **66,** 375–400 (1925).

13. Power, M. H., and Young, N. F., Inorganic phosphate. *In* "Standard Methods of Clinical Chemistry" (M. Reiner, ed.), Vol. I, pp. 84–87. Academic Press, New York, 1953.

14. Woodard, H. Q., A note on the inactivation by heat of acid glycerophosphatase in alkaline solution. *J. Urol.* **65,** 688–690 (1951).

15. King, E. J., and Armstrong, A. R., A convenient method for determining serum and bile phosphatase activity. *Can. Med. Assoc. J.* **31,** 376–381 (1934).

16. Carr, J. J., and Reiner, M., Alkaline and acid phosphatase. *In* "Standard Methods of Clinical Chemistry," (M. Reiner, ed.), Vol. I, pp. 75–83. Academic Press, New York, 1953.

17. Bessey, O. A., Lowry, O. H., and Brock, M. J., A method for the rapid determination of alkaline phosphatase with five cubic millimeters of serum. *J. Biol. Chem.* **164,** 321–329 (1946).

18. Hudson, P. B., Brendler, H., and Scott, W. W., A simple method for the determination of serum acid phosphatase. *J. Urol.* **58,** 89–92 (1947).

19. Huggins, C., and Talalay, P., Sodium phenolphthalein phosphate as a substrate for phosphatase tests. *J. Biol. Chem.* **159,** 399–410 (1945).

20. Seligman, A. M., Chauncey, H. H., Nachlas, M. M., Manheimer, L. H., and Ravin, H. A., The colorimetric determination of phosphatases in human serum. *J. Biol. Chem.* **190,** 7–15 (1951).

21. Bodansky, M., and Bodansky, O., "Biochemistry of Disease," 2nd ed., pp. 717–823. Macmillan, New York, 1952.

22. Hoffman, W. S., "The Biochemistry of Clinical Medicine," pp. 418–473. Year Book, Chicago, Illinois, 1954.

23. Bodansky, M., and Bodansky, O., "Biochemistry of Disease," 2nd ed., pp. 419–421. Macmillan, New York, 1952.

24. Hoffman, W. S., "The Biochemistry of Clinical Medicine," pp. 324–325. Year Book, Chicago, Illinois, 1954.

25. Reinhold, J. G., Chemical evaluation of the functions of the liver. *Clin. Chem.* **1**, 351–421 (1955).
26. Mendelsohn, M. L., and Bodansky, O., The value of liver function tests in the diagnosis of intrahepatic metastases in nonicteric patient. *Cancer* **5**, 1–8 (1952).
27. Gutman, A. B., and Gutman, E. B., "Acid" phosphatase activity of the serum of normal human subjects. *Proc. Soc. Exptl. Biol. Med.* **38**, 470–473 (1938).
28. Gutman, A. B., and Gutman, E. B., An "acid" phosphatase occurring in the serum of patients with metastasizing carcinoma of the prostate gland. *J. Clin. Invest.* **17**, 473–478 (1938).
29. Gutman, A. B., Gutman, E. B., and Robinson, J. N., Determination of serum "acid" phosphatase activity in differentiating skeletal metastases secondary to prostatic carcinoma from Paget's disease of bone. *Am. J. Cancer* **38**, 103–108 (1940).
30. Bodansky, M., and Bodansky, O., "Biochemistry of Disease," 2nd ed., pp. 282–290. Macmillan, New York, 1952.
31. Abul-Fadl, M. A. M., and King, E. J., Properties of the acid phosphatases of erythrocytes and of the human prostate gland. *Biochem. J.* **45**, 51–60 (1949).
32. Fishman, W. H., Dart, R. M., Bonner, C. D., Leadbetter, W. F., Lerner, F., and Homburger, F., A new method of estimating serum acid phosphatase of prostatic origin applied to the clinical investigation of cancer of the prostate. *J. Clin. Invest.* **32**, 1034–1044 (1953).
33. Fishman, W. H., and Lerner, F., A method for estimating serum acid phosphatase of prostatic origin. *J. Biol. Chem.* **200**, 89–97 (1953).
34. Fishman, W. H., Bonner, C. D., and Homburger, F., Serum "prostatic" acid phosphatase and cancer of the prostate. *New Engl. J. Med.* **255**, 925–933 (1956).

PHOSPHATIDES IN PLASMA*

Submitted by: D. B. Zilversmit, Department of Physiology, University of Tennessee, Memphis, Tennessee

Checked by: Myra Janke and W. E. Cornatzer, Department of Biochemistry, North Dakota Medical School, Grand Forks, North Dakota

Margaret Kaser, Veterans Administration Center, Wood, Wisconsin

Introduction and Principles

The phosphatides are complex lipids which contain glycerol (glycerophosphatides, e.g. lecithin and cephalin) or sphingosine (sphingomyelin). All of them are phosphate esters. The hydrocarbon moieties of the molecule account for their solubility in most fat solvents, but acetone, which is a good solvent for triglycerides, precipitates phosphatides. Plasma phosphatides are protein-bound and are precipitated by various protein precipitants. The protein-binding of the phosphatides probably accounts for the fact that the plasma phosphatides cannot be extracted with nonpolar liquids without preceding dehydration, denaturation, or disruption of the protein links. Since the phosphate moiety of the phosphatide molecule is easily determined quantitatively, it is customary to separate the plasma phosphatides from other plasma phosphates and to determine the phosphorus of the lipid fraction. This separation may be accomplished by (a) extraction of plasma phosphatides by a selective fat solvent (ethanol-ether, chloroform-methanol), or (b) the precipitation of plasma phosphatides and their separation from other plasma phosphates. In the following procedure the second approach is taken.

Reagents

1. Trichloroacetic acid, 10%. Make 100 g. of trichloroacetic acid crystals to 1 l. with water. This solution keeps well when stored in a cool place.

* Based on the method of Zilversmit and Davis (1).

2. Perchloric acid, 60% (reagent grade).

3. Ammonium molybdate, 4%. Dissolve 4 g. of finely powdered ammonium molybdate (reagent grade) in 100 ml. water. This solution keeps for several weeks in the refrigerator.

4. 1-Amino-2-naphthol-4-sulfonic acid (A.N.S.A), stock solution. Mix 30 g. sodium bisulfite, 6 g. sodium sulfite, and 0.5 g. of 1-amino-2-naphthol-4-sulfonic acid (Eastman); make to 250 ml. with water. After 2–3 hours remove any undissolved material by filtration and store the clear solution in the refrigerator in a dark bottle. This solution will keep for a month or more. For each series of determinations, 10 ml. of the stock solution is diluted to 25 ml. with water and the solution is then ready for use.

5. Standard phosphate, stock solution: Use 4.39 g. of potassium dihydrogen phosphate per liter. This solution contains 1.000 mg. of phosphorus per milliliter. Add 1 drop $CHCl_3$ as a preservative. In this method use 2.00 ml. of a 1:100 dilution (0.0100 mg. P per milliliter) of the stock solution. Both solutions will keep almost indefinitely in a refrigerator.

Apparatus

1. Micro-Kjeldahl digestion rack. A slightly dented piece of wire gauze supports the digestion tubes. An electric heater at 180°–190°C. is also quite satisfactory.

2. Carborundum boiling chips. Boil with 60% perchloric acid and wash with water before their first use. Hengar granules may also be used instead.[1]

3. Digestion tubes. Pyrex test tubes calibrated to contain 10 ml. Blood sugar tubes are also satisfactory.[1] Some caution should be observed in the use of detergents in washing these tubes since many cleaners contain considerable amounts of phosphate.

4. A photoelectric colorimeter with red filter (maximum transmission at 660 mμ.) A spectrophotometer set at the same wavelength may be used instead.

Procedure

Deliver accurately 0.2 ml. of plasma into a digestion tube containing 3 ml. of water. Add 3 ml. of 10% trichloroacetic acid—the first

[1] Personal communication, Dr. M. M. Best, Department of Medicine, University of Louisville.

1.5 ml. drop by drop while swirling the tube, the remainder more rapidly. Allow the tubes to stand 1–2 minutes, then centrifuge for several minutes until the precipitate is tightly packed; decant the supernatants and invert the tubes until practically all the supernatant has been removed.

Add 1 ml. of 60% perchloric acid to each tube and place a boiling chip in each to insure smooth digestion.

NOTE: Perchloric acid is potentially explosive when hot and in contact with organic matter. It is advisable to watch all digestions until the foaming phase has passed; this requires about 5 minutes.

After 20–30 minutes of heating, the solutions are clear and colorless. When they have cooled, add approximately 5 or 6 ml. of water.

During the digestion, prepare a reagent blank containing 0.8 ml. of 60% perchloric acid and 3 standards containing 0.8 ml. of 60% perchloric acid and 2.00 ml. of the phosphate standards containing 0.010 mg. phosphorus per milliliter. Dilute the blanks and standards with water to a volume of 6 or 7 ml.

NOTE: Only 0.8 ml. of perchloric acid is used because as much as 0.2 ml. of perchloric acid may be used up during the digestion.

Arrange the samples in the following order: the reagent blank, two standards, up to twenty unknowns, one standard. Add 1 ml. of 4% ammonium molybdate to each sample and allow to mix; then add 1 ml. of the ANSA reagent along with enough water to make 10 ml. and again mix the solution thoroughly. Add the ANSA reagent to the samples in the order in which they will be read and note the time of addition. The analyst should standardize his work so that the ANSA reagent is added at approximately the same rate at which the samples are to be read in the colorimeter—approximately 20 minutes after the addition of the ANSA reagent.

Set the colorimeter to zero absorbance with the reagent blank.

Calculation

The absorbance is directly proportional to the phosphorus content of the sample, thus:

$$\frac{\text{Absorbance of unknown}}{\text{Absorbance of standard}} \times 0.02 \times \frac{1}{0.2} \times 100$$

$$= \text{milligrams of phospholipid phosphorus per 100 ml.}$$

NOTE: To compare this value with results expressed as milligrams per cent phospholipid, multiply by 25.

Possible Modifications

If one wishes to determine cholesterol or neutral fat on a plasma sample, in addition to phosphatide, it may be more convenient to prepare one extract of the plasma and determine all three constituents on the same extract. One then determines the total phosphorus content of an ethanol-ether (3:1) extract (2) or on a chlorform-methanol (2:1) extract (3) by evaporating the solvent to dryness on a boiling water bath or electric heater and adding the perchloric acid after the *complete* removal of the solvent. The evaporation of solvent and subsequent digestion may be performed in a test tube, as before, but it has been found convenient to use a 30-ml. Kjeldahl digestion flask. At the end of the digestion one transfers the digest to a 10-ml. volumetric flask or a calibrated test tube, or, to avoid losses by transfer, one adds to the digestion flask 7 ml. of water, 1 ml. of ammonium molybdate, and 1 ml. of ANSA. The exact amount of water added is not very important as long as the same total volume is achieved in standards and unknowns. For a large series of samples the addition of water and reagents may be done with automatic or syringe pipets.

Even though perchloric acid has been used in many laboratories without accidents, some analysts do not like to use this oxidizing agent. It has been shown that other oxidizing agents (H_2SO_4 followed by H_2O_2, or a H_2SO_4 and $HClO_4$ mixture) may be used satisfactorily. In order to minimize the amount of oxidizable material in the digestion tube, it has been found that the plasma may be precipitated with $HClO_4$ instead of trichloroacetic acid so as to give a final acid concentration of approximately 10% (w/v).[2] Objections have been raised to the use of ANSA as a reducing agent because the reaction does not appear to reach a definite end point. The color must therefore be determined at a certain time interval after addition of the reagent. Elon (p-methylaminophenol sulfate) has been recommended in a previous article in this series (4) and has been shown to work satisfactorily for the determination of lipid phosphorus (5). Comparison of Elon and ANSA in the author's laboratory did not show decided superiority of one reagent above the other.

[2] Personal communication from Dr. A. Kaplan, Department of Biochemistry, Michael Reese Hospital, Chicago, Illinois.

Range of Values

The lipid phosphorus concentration in the plasma of normal adults in the postabsorptive state has been found to vary from 6.1 to 14.5 mg. % with a mean of 9.2 mg. % (6). Infants and children before the onset of puberty exhibit plasma phosphatide concentrations as low as one-fourth to one-half the adult values. A meal of moderate fat content does not appear to affect plasma phosphatide levels significantly [pp. 477–478 of reference (6)].

REFERENCES

1. Zilversmit, D. B., and Davis, A. K., Microdetermination of plasma phospholipids by trichloroacetic acid precipitation. *J. Lab. Clin. Med.* **35,** 155–160 (1950).
2. Wittcoff, H., "The Phosphatides," pp. 147–153. Reinhold, New York, 1951.
3. Sperry, W. M., Lipide Analysis. *In* "Methods of Biochemical Analysis" (D. Glick, ed.), p. 106. Interscience, New York, 1955.
4. Power, M. H., Inorganic phosphate. *In* "Standard Methods of Clinical Chemistry" (M. Reiner, ed.), Vol. 1, pp. 84–87. Academic Press, New York, 1953.
5. Maclay, Elizabeth, The determination of lipid phosphorus. *Am. J. Med. Technol.* **17,** 265–270 (1951).
6. Peters, J. P., and Van Slyke, D. D., "Quantitative Clinical Chemistry," 2nd ed., Vol. 1, p. 469. Williams & Wilkins, Baltimore, Maryland, 1946.

PORPHYRINS IN URINE*

Submitted by: ELLEN L. TALMAN, University of Oregon Medical School, Portland, Oregon

Checked by: SAMUEL SCHWARTZ, University of Minnesota Medical School, Minneapolis, Minnesota

Introduction

The methods described here represent minor modifications of those developed by Schwartz and associates (1, 2) which utilize the red fluorescence emitted by porphyrins on exposure to ultraviolet light in the analysis. Coproporphyrin and preformed uro-type (ethyl acetate-insoluble) porphyrins may be determined in a single aliquot of urine. Also, the same procedure is used to separate uro-type porphyrins from residual coproporphyrin after conversion of porphobilinogen and other porphyrin precursors to uroporphyrin by boiling in weakly acid solution. Therefore, the methods for determining both types of compound are described concurrently.

The determination of coproporphyrin is based upon the fact that it is quantitatively extracted from aqueous acetic acid solutions with ethyl acetate. In his original description of the method (1), Schwartz recommended treating the urine with an equal volume of an acetate buffer, composed of 4 volumes of glacial acetic acid and 1 volume of saturated aqueous sodium acetate, prior to extraction with ethyl acetate. Schwartz noted in his paper that uro-type porphyrins are also extracted by ethyl acetate under these conditions. Examination of the buffer revealed that its pH is about 3.1, which is ideal for the extraction of uroporphyrin by ethyl acetate (3). Although these uro-type porphyrins are extracted by the subsequent sodium acetate washes employed, numerous washes are required to effect their complete removal when urines containing large amounts of them (e.g. porphyria urines) are analyzed. Not only is this time-consuming and wasteful of reagents, but it also leads to large and

* Based on the method of Schwartz *et al.* (1, 2).

troublesome volumes of aqueous phase. Accordingly, the buffer described below (pH 4.8) has been adopted in this laboratory. The pH of this buffer lies well within the pH range (4 to 6) at which coproporphyrin can be extracted from aqueous acetic acid solutions by ethyl acetate. At this pH uro-type porphyrins remain in the aqueous phase. Recovery of added coproporphyrin is 98%. Measurement of "uroporphyrin" (i.e. ethyl acetate-insoluble porphyrins), as determined in the aqueous phase and the sodium acetate washes, is remarkably reproducible.

Principle

Coproporphyrin is quantitatively extracted from aqueous acetic acid, pH 4–6, with ethyl acetate. After appropriate treatment of the ethyl acetate phase, the coproporphyrin is extracted from it with 1.5 N HCl, and is then measured fluorometrically.

At pH 4–6, uro-type porphyrins, which remain in the aqueous phase, are adsorbed on aluminum oxide. Following appropriate washing of the aluminum oxide, the porphyrins are eluted with 1.5 N HCl and measured fluorometrically.

In urine uncontaminated by feces, the ethyl acetate-soluble porphyrin is principally coproporphyrin. In urine contaminated by feces or in other biological fluids such as bile or the allantoic fluids from embryonated eggs—to which this method has been successfully applied in this laboratory—the ethyl acetate-soluble fraction is composed of a mixture of porphyrins, principally copro- and protoporphyrins. As indicated above, the porphyrins which are soluble in water but insoluble in ethyl acetate at pH 4–6 are also a mixture, rather than pure uroporphyrin bearing 8 carboxyl groups. For convenience, however, the two fractions are denoted as "coproporphyrin" and "uroporphyrin."

Reagents

1. Ethyl acetate, reagent grade.

NOTE: This solvent may be reclaimed as follows: Wash with 10% sodium carbonate until the aqueous phase remains alkaline, then with water until the aqueous phase is neutral or slightly acid. Dry over calcium chloride and distill from fresh calcium chloride (1). If it is desirable to delay recovery of the used ethyl acetate in this way soon after use, it is advisable to store it over dry sodium carbonate until the recovery process can be carried out. This prevents extensive hydrolysis of the solvent by the acid it contains. One lot of ethyl acetate received in this laboratory, even though of reagent grade, contained an unknown impurity which interfered with the determination. This impurity also carried through the reclamation process.

2. Sodium acetate, saturated aqueous. Saturation is greatly facilitated by dissolving in hot (60°–70°C.) water. If sodium acetate does not crystallize on cooling, warm the solution again and add more of the salt. The anhydrous form or the trihydrate may be used.

3. Sodium acetate, half-saturated. Mix equal volumes of saturated sodium acetate and water.

4. Sodium acetate, 1%. Dissolve 10 g. of anhydrous CH_3COONa or 16.6 g. of $CH_3COONa.3H_2O$ in water and dilute to 1 l. This solution becomes moldy rather quickly.

5. Acetic acid, glacial, reagent grade.

6. Acetate buffer, pH 4.8. Mix 1 volume of glacial acetic acid, 4 volumes of saturated sodium acetate and 3 volumes of water.

7. Iodine stock solution, 1% in alcohol. Dissolve 1 g. of iodine crystals (U.S.P.) in ethanol and dilute to 100 ml. Store in a brown, glass-stoppered bottle in the refrigerator.

8. Iodine, 0.005% aqueous. Dilute 1 ml. of the stock 1% iodine in alcohol to 200 ml. with water. Prepare fresh daily.

9. Hydrochloric acid, 1.5 N. Dilute 125 ml. of concentrated HCl (reagent grade) to 1 l. with water.

10. Coproporphyrin stock standard. Dissolve 0.542 mg. of crystalline coproporphyrin tetramethyl ester in 1.5 N HCl, then dilute to 100 ml. with the same solvent. Let stand 4–5 hours to insure complete hydrolysis. This solution contains 500 μg. of coproporphyrin per 100 ml. Store in a dark, glass-stoppered bottle in the refrigerator.

NOTE: The checker believes that complete solution of the standard is achieved more readily and reliably in 7.5 N HCl.

Coproporphyrin for the preparation of standards is obtained by purification of the compound isolated from urine (preferably porphyria urine, since it contains such large amounts). The primary isolation may be accomplished by a technique analogous to that employed in the analytical procedure. In this case, after adjustment of the urine to pH 4–6 with acetic acid and/or acetate buffer, an equal volume of ethyl acetate is used to extract the coproporphyrin. After the coproporphyrin has been extracted from the ethyl acetate phase with 1.5 N HCl, its tetramethyl ester is prepared by adding 10 volumes of $CH_3OH:H_2SO_4$ (20:1) [see reference (4)]. The ester is then separated from the reaction mixture, chromatographed, and crystallized as described by Grinstein *et al.* (5). Coproporphyrin obtained commercially is likely to require similar purification. Before it is used as a standard, the preparation should be checked for purity by comparing some of its physical constants with those reported in the literature for coproporphyrin tetramethyl ester. The absorption spectrum and melting point are the criteria most commonly used. These data have been recorded by Lemberg and Legge (6). A similar comparison with the standard material prepared by other investigators is also desirable.

NOTE: The checker prefers to neutralize the HCl solution of porphyrin with cold NaOH to pH 3–6, extract with ethyl acetate, wash the ethyl acetate with water, and vacuum distill to dryness. The porphyrin residue is esterified with $CH_3OH:H_2SO_4$ (20:1).

Coproporphyrin Working Standards

Diluting 1.00 ml. of the stock standard coproporphyrin to 100 ml. with 1.5 N HCl gives a working standard containing 5.00 µg. per 100 ml. Although the relationship between concentration and fluorescence is linear up to 5.00 µg. per 100 ml., it is desirable to have available more dilute standards for the precise determination of very small quantities of porphyrins. These are easily prepared by making appropriate dilutions of the stock standard with 1.5 N HCl. Unknowns containing more than 5.00 µg. per 100 ml. are diluted for comparison with the standards.

For general use, transfer an appropriate volume of each working standard to a fluorometer test tube cuvette, stopper, and seal with paraffin. Standards prepared in this way are quite stable even when kept at room temperature and used several times a day. When not in use, they should be stored in a dark place. Also, they should be checked periodically against standards stored in the refrigerator.

Special Equipment†

1. Ultraviolet lamp. This is required to check 1.5 N HCl extracts for red fluorescence visible in the dark. Maximum fluorescence is

† The authors will gladly supply more information regarding specific items of equipment and brands of reagents upon request.

excited by the 405 mμ light from a mercury-arc lamp. A lamp with a focusing beam is desirable.

2. Fluorometer. A sensitive fluorometer, usually one including a photomultiplier tube in its design, should be available if these methods are to be utilized. Several fluorometers are now available. Schwartz (1) has described the construction of suitable apparatus. Several combinations of primary and secondary filters were studied by the Schwartz group (1) during the development of their method for coproporphyrin. Of these combinations, that utilizing a Corning No. 5113 filter (5–58, which absorbs maximally at 405 mμ) for the primary and a Corning No. 2412 filter (2–61, which absorbs maximally at 615 mμ) for the secondary has proved very satisfactory in this laboratory.

Procedure

1. Preservation of urine. Schwartz *et al.* (1) found pH to be a critical factor in the preservation of urinary coproporphyrin and reported a pH range of 6.5 to 9.5 as satisfactory. The importance of pH in the preservation of urinary coproporphyrin has been confirmed in this laboratory, but these studies showed that the pH may drop as low as 6.0 without serious loss of this porphyrin. However, if the pH is allowed to fall below 6.0, marked losses occur. Coproporphyrin is also lost above pH 9.5, but these losses are much smaller than those observed when the urine is too acid. The preservation studies carried out in this laboratory also indicate that the presence of an antibacterial agent, such as toluene or thymol, aids in the preservation of urinary coproporphyrin.

Schwartz *et al.* (1) suggested sodium carbonate, in a final concentration of approximately 0.3–1% (5 g. for a 24-hour urine sample), as a preservative and this treatment serves well in many cases. However, the studies made in this laboratory indicate that a lower sodium carbonate concentration, approximately 0.1%, supplemented by a layer of toluene may be preferable.

Metal caps, or metal inserts in the caps, on urine-collection bottles are to be avoided. If porphyria or liver disease are suspected, make the collections in dark bottles. Analyses should be completed within a week after collection, even if the samples are refrigerated, which is desirable. Since the phosphates which precipitate in alkaline urine may adsorb porphyrins, these urines should never be filtered prior to analysis.

2. *Determination of coproporphyrin.* Transfer an aliquot of well-mixed urine containing 1.25 μg. or less of coproporphyrin to a 250-ml. separatory funnel. (Usually 5.00 ml. of normal urine is satisfactory, but urines containing large quantities of porphyrins should be diluted prior to analysis.) Add 5 ml. of acetate buffer, 10 ml. of water, and 75–100 ml. of ethyl acetate. Shake thoroughly, then allow the two phases to separate. Drain the aqueous phase into a 50-ml., conical-tip centrifuge tube containing approximately 0.5 g. of aluminum oxide, mix, and hold for estimation of preformed uroporphyrin. Wash the ethyl acetate phase with two 10-ml. portions of 1% sodium acetate or preferably until these washes show no red fluorescence, combining these washes with the original aqueous phase in the centrifuge tube. If the urine sample is fresh, add 5–10 ml. of 0.005% aqueous iodine to the ethyl acetate phase, mix very gently, draw off and discard the aqueous phase. If an emulsion forms, it can usually be broken by adding a little ethanol. Limit iodine action to 5 minutes, if possible. Omit the iodine treatment if the sample has been standing for 24 hours or more after the collection was completed. Extract coproporphyrin from the ethyl acetate with four 5-ml. portions of 1.5 N HCl. If the fourth HCl extract exhibits red fluorescence in the dark upon exposure to ultraviolet light (405 mμ), extraction must be repeated until such fluorescence is no longer visible. Dilute the combined extracts to 25.0 ml., or some other convenient volume, with 1.5 N IICl and read in the fluorometer, which is standardized with the pure coproporphyrin described above.

NOTE: The checker prefers to add the iodine to the first ethyl acetate extract.

3. *Determination of preformed uro-type porphyrins.* Mix the aluminum oxide thoroughly with the aqueous phase from the coproporphyrin determination and sediment the aluminum oxide by centrifugation at 1500–2000 r.p.m. Discard the supernatant. Resuspend the aluminum oxide in 20 ml. of half-saturated sodium acetate and centrifuge as before. Wash with 40 ml. of water in a similar manner. Discard each supernatant. Elute uroporphyrin from the aluminum oxide by resuspending the adsorbent in 5 ml. of 1.5 N HCl, and sedimenting as before. Repeat this step three times. If

the supernatant from the fourth elution exhibits any red fluorescence in the dark upon exposure to 405 mμ ultraviolet light, elution must be repeated until this fluorescence disappears. Dilute the combined eluates to 25.0 ml., or some other suitable volume, with 1.5 N HCl and read in the fluorometer against a pure coproporphyrin standard. Coproporphyrin standards are used for reference because uroporphyrin standards do not keep well. The coproporphyrin standards may be calibrated against freshly prepared solutions of uroporphyrin, or the uroporphyrin isolated from urine can be read directly against the coproporphyrin standards. The latter procedure is quite satisfactory for comparative purposes.

NOTE: The checker recommends applying a factor of 0.75 to the uroporphyrin value obtained when comparing uroporphyrin with coproporphyrin standards using the primary and secondary filters described.

This technique for estimating uroporphyrin yields consistently reproducible results, but further investigation of its accuracy and limitations would be desirable.

4. Determination of total uroporphyrin. To convert uroporphyrin precursors to uroporphyrin, acidify an aliquot of urine to pH 5.0 with glacial acetic acid and heat in a boiling water bath for 30 minutes. Transfer quantitatively to a separatory funnel and separate uroporphyrin from residual coproporphyrin as outlined above. Coproporphyrin cannot be determined in samples treated in this way, however, since much of the coproporphyrin is destroyed by heating.

Calculations

$$\frac{\text{F.R. } u}{\text{F.R. } s} \times S \times V \times \frac{1}{A} \times D = \mu\text{g. Porphyrin/100 ml. urine}$$

where F.R.u = fluorometer reading of the unknown.

F.R.s = fluorometer reading of the standard.

S = concentration of the coproporphyrin standard in μg./100ml.

V = volume of the combined HCl extracts.

A = aliquot of sample taken for analysis.

D = dilution, either of the urine or of the final extract.

If a 5.00-ml. aliquot of urine is used and the volume of the combined extracts is 25.0 ml., this equation may be simplified to:

$$\frac{\text{F.R. } u}{\text{F.R. } s} \times S \times 5 \times D = \mu\text{g. Porphyrin}/100 \text{ ml. urine}$$

$$\mu\text{g. Porphyrin per day} = \frac{\mu\text{g. Porphyrin}/100 \text{ ml.}}{100}$$

$$\times \text{ 24-hr. urine volume (in milliliters)}$$

Discussion

Since porphyrin fluorescence decreases with increasing HCl concentration, the acidity of the porphyrin-containing extracts must be controlled as described above.

Saturation of HCl solutions of porphyrins with ethyl acetate increases coproporphyrin fluorescence by approximately 8% over the pure HCl solution (1). For very precise work, it may be desirable to increase the concentrations of the standards by that amount.

The presence of colored impurities may reduce porphyrin fluorescence. Therefore, if the extract is colored, it should be diluted until essentially colorless before reading in the fluorometer.

Approximately half of the porphyrin in freshly voided urine is present in the form of a nonfluorescent precursor (1). Conversion of the coproporphyrin precursor to the fluorescent porphyrin is accomplished by washing with 0.005% iodine or by allowing the urine to stand for 24 hours before analysis. With freshly passed urine, to use too much iodine or to allow the iodine to act for too long a time can result in the loss of porphyrin. If the urine is stored for at least 24 hours before analysis, iodine treatment may not increase the amount of porphyrin detected and may slightly decrease porphyrin content, even when the amount of iodine used and its time of action are carefully controlled. Thus, it is advisable to exercise considerable caution in the use of iodine with fresh samples and to omit this step when analyzing older urines. The occurrence of these precursors in various biological fluids and their conversion to the porphyrins have been discussed thoroughly by Watson and associates (7).

In the course of this procedure, solid materials, which are not removed by the sodium acetate washes, sometimes precipitate. If this material collects at the interphase, as it often does, it is not difficult to exclude it from the extracts. However, if it does not collect at the interphase, the precipitate may be removed by draining the acid extracts into the receiver through a plug of Pyrex glass wool placed in the stem of a small funnel. If such filtration is necessary, the funnel and plug must be carefully washed free of porphyrin.

Range of Values in Healthy Persons

Urinary coproporphyrin excretion is directly related to body weight (8, 9). Although it is now known that traces of uroporphyrin occur in normal human urine (10), little quantitative data is available to date. The values tabulated in Table I were obtained using the methods described above. Figures for adults have not been broken down according to body weight.

Pathological States

Large quantities of coproporphyrin and uroporphyrin are excreted in both hepatic and erythropoietic porphyria. Excessive coproporphyrinuria occurs in a number of other disorders including lead poisoning, rheumatic fever, poliomyelitis, infectious hepatitis, cirrhosis, and some of the anemias (10), and has been reported in a few cases of malignant neoplastic disease as well (13). Urinary uroporphyrin has not been studied in most of these diseases. However, Watson has observed uroporphyrinuria in some of his patients suffering from malignancies (13).

TABLE I

URINARY PORPHYRINS

	Coproporphyrin (μg./day)		Uroporphyrin (μg./day)	
	Range	Mean	Range	Mean
Adults:				
This laboratory[a]	81.0–222	132	28.1–63.4	39.0
Schwartz (1)	—	160	—	—
Bashaur (11)	—	—	2–37	—
Lockwood (12)[b]	—	—	11–40	—
Children (8):				
61–80 lb.	50–119	75.6	—	—
81–100 lb.	50–165	94.1	—	—
101–110 lb.	76–172	113	—	—

[a] Figures derived from 9 healthy subjects.
[b] Data from 6 healthy adults.

REFERENCES

1. Schwartz, S., Zieve, L., and Watson, C. J., An improved method for the determination of urinary coproporphyrin and evaluation of the factors influencing the analysis. *J. Lab. Clin. Med.* **37**, 843–859 (1951).
2. Schwartz, S., Keprios, M., and Schmid, R., Experimental porphyria II. Type produced by lead, phenylhydrazine and light. *Proc. Soc. Exptl. Biol. Med.* **79**, 463–468 (1952).
3. Dresel, E. I. B., and Tooth, B. E., Solubility of uroprophyrin I in ethyl acetate, *Nature* **174**, 271 (1954).
4. Schwartz, S., Hawkinson, V., Cohen, S., and Watson, C. J., A micromethod for the quantitative determination of the urinary coproporphyrin isomers (I and III). *J. Biol. Chem.* **168**, 133–144 (1947).
5. Grinstein, M., Schwartz, S., and Watson, C. J. Studies of the uroporphyrins. I. The purification of uroporphyrin I and the nature of Waldenström's uroporphyrin, as isolated from porphyria material. *J. Biol. Chem.* **157**, 323–344 (1945).
6. Lemberg, R., and Legge, J. W., "Hematin Compounds and Bile Pigments," pp. 63–65, 74. Interscience, New York, 1949.
7. Watson, C. J., Pimenta de Mello, R., Schwartz, S., Hawkinson, V., and Bossenmaier, I., Porphyrin chromogens or precursors in urine, blood, bile, and feces. *J. Lab. Clin. Med.* **37**, 831–842 (1951).
8. Strait, L. A., Bierman, H. R., Eddy, B., Hrenoff, M., and Eiler, J. A., Relation of body weight and urinary coproporphyrin excretion. *J. Appl. Physiol.* **4**, 699–708 (1952).
9. Neve, R. A., and Aldrich, R. A., Porphyrin metabolism. III. Urinary and erythrocyte porphyrin in children with acute rheumatic fever. *Pediatrics* **15**, 553–561 (1955).
10. Aldrich, R. A., Labbe, R. F., and Talman, E. L., A review of porphyrin metabolism with special reference to childhood. *Am. J. Med. Sci.* **230**, 675–697 (1955).
11. Bashaur, F., The study of uroporphyrin metabolism with special reference to conditions other than porphyria. Ph.D. Thesis, University of Minnesota, Minneapolis, Minn., 1956.
12. Lockwood, W. H., Uroporphyrins: I. Uroporphyrin content of normal urine. *Australian J. Exptl. Biol. Med. Sci.* **31**, 453 (1953).
13. Watson, C. J., Some studies of nature and clinical significance of porphobilinogen. *A.M.A. Arch. Internal Med.* **93**, 643–657 (1954).

PROTEIN-BOUND IODINE IN SERUM*

Submitted by: George R. Kingsley Departmentof Physiological Chemistry, School of Medicine, University of California at Los Angeles and Veterans Administration Center, Los Angeles, California, and Roscoe R. Schaffert, Veterans Administration Center, Los Angeles, California.

Checked by: Robert L. Dryer, University Hospitals, State University of Iowa, Iowa City, Iowa

Joseph V. Princiotto, Georgetown University School of Medicine, Washington, D. C.

Introduction

The use of iodine salts to catalyze the reduction of cerium salts by the arsenite ion suggested by Kolthoff and Sandell (1) has been generally adopted as the most sensitive method for the detection of protein-bound iodine (PBI) in 1–2 ml. of serum. The two most commonly used methods for separation of the PBI from serum for application of the cerium-arsenite-iodide reduction, are the sulfuric-chromic acid digestion and distillation method of Chaney (2) and the alkaline-ash method of Barker and Humphrey (3). Critical evaluation of the acid digestion-distillation methods for the determination of serum PBI has been made (4, 5). The original Chaney still has been further modified (6, 7) and a simple retort type of still proposed (8). Elimination of distillation of iodine from the acid digest has been suggested (9–11) by the use of chloric acid digestion, which, however, requires an increase in the ratio of arsenious to ceric reagent (10:1) as a result of the presence of chromium and other interfering ions. Techniques have been described to slow up the reduction of the cerium salts with mercury (12) and to stop the reaction with brucine in order to have more time to measure optical density (13). A brief historical survey of blood iodine methods has been made (14).

* Based on the methods of Chaney (2) and Barker and Humphrey (3).

147

State of Iodine in Serum

Iodine is present in blood serum in two forms "organic" and "inorganic." The organic is "protein-bound" in the principal hormonal forms, thyroxine and triiodothyronine. The concentration of the "organic" form gives an indication of thyroid activity. The "inorganic" iodine present in serum does not correlate with thyroid activity but indicates the balance between intake, metabolism, and excretion of iodine. The inorganic iodine for this reason must be separated from the organic when the concentration of the latter is used as a measure of thyroid activity.

Preparation of Patient Prior to Collection of Sample

The patient must not have received an inorganic iodine medication for 24–48 hours prior to the test. Iodine application to the skin, iodine therapy with Lugol's solution, or other medication must be avoided. Unfortunately, exposure of serum proteins to abnormal amounts of inorganic iodide resulting from excessive absorption may result in binding of the iodine to serum proteins, in which state the iodine cannot be separated by washing. Diagnostic procedures employing organic iodine compounds may be divided into three general groups as to the length of time in which they will interfere with measurement of PBI: (a) intravenous pyelograms—up to 1 month; (b) gall bladder dyes and therapeutic drugs such as Itrumil, Diodrast, and Diodoquin—2–6 months; (c) Lipiodol and other compounds administered intrathecally—6 months to 5 years (15). Thyroxine and thiouracil will interfere for about 1 month with serum PBI base-level measurement. Mercurial compounds such as those used as diuretics will cause falsely low serum PBI values for as long as 2 days if the acid digestion—distillation method is used but does not interfere with the dry-ash method. Diets rich in sea food may increase the serum PBI.

Collection of Samples

A 12–15-ml. sample of blood is usually sufficient for performing the test in duplicate. Syringe, needle, and test tube used in collecting blood for PBI test should be chemically clean and kept wrapped or covered. After the specimen has been allowed to clot for 30 minutes or longer in a stoppered test tube, the serum is separated by centrifugation and may be kept for several days at room temperature or as long as protein remains soluble without change in PBI content. Serum containing 0.2% benzoic acid has been kept 15 days at 20–25°C. without alteration of PBI. Fasting specimens are not required. A resting state is also not required, as moderate exercise does not alter the PBI tests.

Principle of the Method

The proteins of serum are precipitated with sulfuric acid-zinc sulfate and potassium hydroxide. The proteins are washed free of unbound iodine. The protein-bound iodine is converted to inorganic iodine by either of two methods (a) acid digestion or (b) by alkaline dry ash incineration. In (a) the serum proteins are digested directly with a mixture of sulfuric and chromic acids to convert the iodine to inorganic iodides. The iodine in the reduced state is distilled off, after the addition of phosphorous acid, into an absorbant of potassium hydroxide or sodium arsenite. In (b) the proteins are dried and incinerated at 600°C. with potassium carbonate to remove all organic material and convert the iodine to inorganic salts. The amount of inorganic iodine formed in either of the above digestion methods is determined photometrically by its catalytic action on the rate of reduction of yellow ceric ion to colorless cerous ion in the presence of arsenite ion.

I. DISTILLATION METHOD

Glassware

All glassware stills, pipets, etc. used in the iodine procedure must be carefully washed with sulfuric—chromic acid reagent or a satisfactory detergent as shown by actual test. (Some commercial detergents have been found to be satisfactory.) It may be preferable to rinse with dilute KOH after the acid wash. This glassware must be kept segregated from other glassware. Stopcocks may be lubricated with some tested lubricants if great care is exercised and very little is used (avoid silicone-base lubricants). Iodine determinations are generally more satisfactory if carried out in a special area or room free of chemical dust, fumes, formaldehyde, iodine, mercury, etc.

Iodine still: See Fig. 1 for specifications of the modified Chaney still.

Reagents

All reagents are stored in acid-washed Pyrex, glass-stoppered containers. All reagents must be prepared with redistilled water. The total blank of all reagents should contain less than 0.01 μg. of iodine (in a 5-ml. aliquot of distillate).

Distilled water: Redistill a good grade of distilled water (some commercial distilled water has been found to be satisfactory) containing 5 ml. of 0.25 N KOH per liter of water in an all-glass still.

1. 27 N Sulfuric acid. Prepare 50% (by volume) sulfuric acid from reagent grade concentrated H_2SO_4 and redistilled water. Place it in an all-glass still and distill until 27 N sulfuric acid is obtained in the distillation flask. Some reagent grades of concentrated sulfuric acid need no further purification, and 27 N sulfuric acid may be prepared by diluting approximately 75 ml. of concentrated H_2SO_4 (sp. gr. 1.84) to 100 ml. with redistilled water.

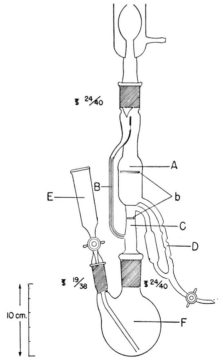

FIG. 1. Modified Chaney Distillation Assembly; A, still head; B, capillary return to still; C, vapor and spray trap; D, distillate trap; E, dropping funnel; F, distillation flask; b, baffle plates.

2. 100% Chromic acid. Dissolve 100 g. of chromic acid in about 60 ml. of redistilled water and dilute to 100 ml. Some U.S.P. or technical grades of chromic acid contain less iodine than the C.P. grade. (Merck's reagent grade has been satisfactory.)

3. *Ceric hydroxide reagent.* (G. Frederick Smith Chemical Company, Columbus, Ohio.) This reagent is more uniform than other ceric reagents.

0.0114 N Ceric hydroxide, Ce(OH)$_4$ (Check by titration.) Dissolve 0.2375 g. of ceric hydroxide in 100 ml. of 1.81 N H$_2$SO$_4$ on a boiling water bath. Cool to room temperature and dilute to 100 ml. with 1.81 N H$_2$SO$_4$. Let stand 30 minutes to clear. Higher concentrations of this reagent (0.05 N) depress the rate of reduction and give a narrow range of iodine measurement (see Figs. 3 and 6). The concentration of the ceric hydroxide reagent may be varied for convenience in the use of the photometer to obtain a blank reading of approximately 20% transmittance.

4. Potassium hydroxide, approximately 14 N. Mix 7 parts by weight of reagent grade potassium hydroxide with 5 parts of redistilled water. Continue mixing while hot until all solid potassium hydroxide is completely dissolved. Cool to room temperature (keep temperature above 20°C.). All potassium hydroxide solutions are prepared from this reagent.

5. 0.25 N Potassium hydroxide. Prepare a 0.25 N solution of potassium hydroxide from 14 N reagent grade KOH with redistilled water.

6. 0.06 N Arsenious acid in 1.8–2.0 N H$_2$SO$_4$. Add 0.2967 g. of arsenious trioxide (As$_2$O$_3$), reagent special, primary standard (Baker and Adamson or equal) to a 100-ml. volumetric flask and dissolve in 8 ml. of 0.75 N KOH (prepared from 14 N KOH). Warm if necessary to dissolve. Add 7.4 ml. of 27 N H$_2$SO$_4$, mix, dilute to 100 ml. with redistilled water at room temperature, and mix again.

NOTE: Investigation has indicated that approximately 0.50 g. of NaCl must be added to this reagent to obtain reduction of ceric reagent to give a good working range of per cent transmittance. (See Fig. 2.)

7. 0.45 N Arsenious acid in 1.3–1.4 N H$_2$SO$_4$. Add 2.2253 g. of arsenious trioxide (see grade above) to a 100-ml. volumetric flask and dissolve in 6 ml. of 7.5 N KOH. Warm if necessary to dissolve.

Add 7.4 ml. of 27 N H_2SO_4, mix, and dilute to 100 ml. with redistilled water at room temperature and mix again.

8. 1% Sodium arsenite ($NaAsO_2$). (*Optional substitute for reagent No. 5.*) Dissolve 1.0 g. of reagent grade sodium arsenite in 100 ml. of redistilled water.

9. Stock standard iodine solution. Dissolve 0.1309 g. of reagent grade KI (correct for indicated moisture content) in 100 ml. of redistilled water; 1 ml. = 1 mg. iodine. Store in a glass-stoppered dark bottle.

10. Dilute stock standard iodine solution. Dilute 1.00 ml. of stock standard iodine solution to 100 ml. with redistilled water and store as directed above; 1 ml. = 10 μg. of iodine.

11. Dilute working standard iodine solution. Dilute 1.00 ml. of dilute stock standard iodine solution to 100 ml. with redistilled water and use when freshly prepared; 1 ml. = 0.1 μg. of iodine.

12. 50% Phosphorous acid. Dissolve 50 g. of reagent grade phosphorous acid (Baker and Adamson, J. T. Baker, or equal) in redistilled water and dilute to 100 ml.

13. Reagents for serum protein precipitation.

a. Zinc sulfate ($ZnSO_4 \cdot 7H_2O$). Add 12.5 g. of reagent grade zinc sulfate to a 1-l. volumetric flask containing 125 ml. of 0.25 N H_2SO_4 (31.25 ml. of 1 N H_2SO_4). Mix, dissolve, and dilute to volume with redistilled water.

b. 0.75 N Potassium hydroxide. Prepare from 14 N iodine-free potassium hydroxide. When titrated, 2.00 ml. should be equivalent to 12.8 ± 0.1 ml. of zinc sulfate solution. If an approximately neutral solution is not obtained (pH 6–8) make proper adjustment of the two protein-precipitating reagents.

14. Iodine protein carrier standard and blank protein carrier reagents. Dissolve 6–7 g. of Armour Bovine Albumin (Armour Laboratories, Chicago, Illinois), Fraction V, in 100 ml. of distilled water. (This reagent should not contain more than 1 μg. of iodine in 100 ml.)

NOTE: All the above reagents, if prepared properly, are stable.

Procedure

PRECIPITATION OF PROTEIN AND WASHING OF PRECIPITATE

Add 2.00 ml. of serum (slight hemolysis is permissible) to 16.0 ml. of zinc sulfate solution in a 28 x 120-mm. Pyrex centrifuge test tube. Mix by rotating and then add 2.00 ml. of 0.75 N KOH. Mix thoroughly and let stand 15 minutes. Centrifuge until the precipitate is well packed and the supernatant fluid is clear. Pour off the supernatant and reserve for the inorganic iodine and contamination tests. Test 5.0 ml. of the supernatant for iodine as described for 5.0 ml. of the distillate under PHOTOMETRIC MEASUREMENT: *Color Development*. Add 20 ml. of redistilled water to the protein precipitate and suspend the precipitate thoroughly by mixing with a glass stirring rod. Centrifuge again as directed above. If the supernatant indicates more than 25 μg. of inorganic iodine/100 ml. in the original serum, a second wash is required. If it is greater than 200 μg./100 ml., contamination is indicated and a satisfactory PBI analysis of the specimen cannot be made.

DIGESTION OF PROTEIN PRECIPITATE

Dissolve the washed precipitate in 15 ml. of 27 N H_2SO_4 and transfer quantitatively to the two-necked digestion flask of the modified Chaney still. Remove the remainder of the dissolved precipitate by washing out the centrifuge tube with two 5.0 ml. portions of 27 N H_2SO_4. Add 2.0 ml. of the 100% chromic acid reagent and two clean glass beads or a few pre-ashed carborundum particles and then digest 7–8 minutes over a 350-watt cone electric heater (Cenco, Chicago, Illinois) in a fume hood or device connected to a manifold connected to a suction pump. Swirl the flask a few times to mix and permit smooth boiling. Continue boiling until the mixture is brown green in color and white SO_3 fumes just appear. If the solution turns green, more chromic acid reagent is required. If a white precipitate forms, the digestion has been carried too far. Allow the flask to cool 5 minutes. Add 15 ml. of redistilled water and boil again until white fumes just appear. Cool, add 25 ml. of redistilled water, and mix.

DISTILLATION OF THE IODINE FROM THE DIGEST

Connect the center outlet of the digestion flask with the trap and place the condenser on top of the trap. Insert the dropping funnel, with stopcock closed, into the side outlet. Assemble the whole apparatus on a 350-watt conical electric heater (Cenco), add 1.0 ml. of 50% phosphorous acid reagent to the dropping funnel. Add 1.0 ml. of 0.25 N KOH to the trap. (1.0 ml. of 1% sodium arsenite may be added to the trap instead of potassium hydroxide.)

Adjust the bottom extended lip of the condenser to permit delivery into the side capillary return to the flask. When the mixture in the flask is boiling, add 1.0 ml. of 50% phosphorous acid reagent slowly from the side funnel. Approximately 15–30 seconds after the addition of the phosphorous acid add 1.0 ml. of 0.45 N arsenious acid (The solution in flask should be bright green in color) and continue distillation 7–8 minutes after the phosphorous acid has been added. Remove the dropping funnel at the end of the distillation and discontinue heating of the distillation flask. Run the contents of the trap into a 25-ml. graduated cylinder or tube. Wash the still trap 3 times from top to bottom with a minimum of redistilled water. The combined volume of distillate and washings should be 12–14 ml. Dilute to 15 ml. with distilled water.

PHOTOMETRIC MEASUREMENT

Preparation of Specimens

(a) *Serum: Preparation of distillates from digests of serum:* Add 5.0-ml. aliquots of the distillate to two photometer tubes.

(b) *Blank: Preparation of distillates from digests of reagents:* Add two 5.0-ml. aliquots of the distillate from the digestion of 2.00 ml. of 6–7% bovine albumin and all reagents (prepared as outlined in procedure for serum) to photometer tubes.

(c) *Preparation of absolute iodine standards:* Add 4.0-ml. aliquots of a blank distillate as described under (b) to each of four photometer tubes containing 0.0, 0.2, 0.5, and 1.0 ml. of the dilute working standard iodine solution (1.0 ml. = 0.1 μg. iodine). Dilute to 5.0 ml. with redistilled water.

(d) *Preparation of recovery iodine standards (with carrier):* Precipitate and wash the protein of four samples of 2.0 ml. of 6–7% bovine albumin. Add 0.0, 0.6, 1.5 and 3.0 ml. of dilute working standard iodine solution (1.0 ml. = 0.1 μg. iodine) to each of the protein precipitates and continue digestion as directed above for serum. Prepare distillates for photometric measurement as directed for serum.

Color Development

Add to the photometer tubes prepared as described above in (a), (b), (c), and (d), 1.0 ml. of 0.06 N arsenious acid reagent and mix by twirling. (Do not invert, stopper, or use fingers.) Place all tubes in a constant-temperature heating block or water bath maintained at a constant temperature of 25°, 30°, or 37°C., selected according to the ceric hydroxide reagent and photometer used. When the temperature of the tubes has reached equilibrium, add exactly 1.00 ml. of ceric hydroxide reagent to each tube at 30-second intervals. Mix as directed above and read in the photometer at an appropriate time (selected according to the temperature used to give proper range of light absorption) after the addition of the ceric hydroxide. Set the photometer with a water blank at 100%T using a wavelength of 420 mµ. Other time intervals may be used if more convenient because of the type of photometer used or the rate of reaction.

Calculation

Plot on rectilinear coordinate or semilog graph paper the per cent transmittance of the absolute iodine standards and reagent blank against the corresponding concentrations of iodine equivalent to micrograms of iodine per 100 ml. of serum. The amount of iodine present in each aliquot of distillate is read directly from this graph. It may be calculated as follows:

$$\text{Micrograms iodine/100 ml.} = \frac{100}{2} \times \frac{15}{5} \times \text{conc. std.}$$

$$\times \frac{\%T \text{ unknown} - \%T \text{ of blank}}{\%T \text{ std.} - \%T \text{ of blank}}$$

(This calculation applies only to straight-line curve graph data.)

Per cent recovery is calculated as follows from the data of the recovery iodine standards:

$$\% \text{ Recovery}$$

$$= \frac{\text{Total iodine recovered (micrograms \%)} - \text{bovine iodine (micrograms \%)}}{\text{Iodine added (micrograms \%)}}$$

$$\times 100$$

FINAL REPORT

$$\text{Corrected PBI} = \text{Serum PBI from absolute iodine chart} \times \frac{100}{\% \text{ recovery}}$$

Testing of Reagents

The reagents should be checked for their iodine content, inhibition, or acceleration effect in the following order and manner:

1. 0.06 N Arsenious acid reagent. Add 1.0 ml. of arsenious acid reagent to a photometer tube and dilute to 5.0 ml. with redistilled water. Add 1.0 ml. of ceric hydroxide reagent and mix. No *appreciable* decolorization should occur within 1 hour after initial mixing. If decolorization occurs equivalent to 1–2 µg. per cent iodine, the arsenious acid reagent is contaminated. The presence of inhibiting materials in this reagent may be checked by the addition of iodine standards.

2. 27 N Sulfuric acid. Dilute 1.0 ml. 0.6 N H_2SO_4 made up from 27 N H_2SO_4 to 5.0 ml. with water, add 1.0 ml. of arsenious acid reagent, and continue as directed under *Color Development*.

3. 0.25 N Potassium hydroxide. Dilute 0.1–0.3 ml. of 0.25 N KOH to 5 ml., with water, add 1.0 ml. of arsenious acid reagent, and continue as directed under *Color Development*.

4. 100% Chromic acid and 50% phosphorous acid. Run through digestion, distillation, and color development as directed in the procedure.

5. Zinc sulfate ($ZnSO_4.7 H_2O$). Add 2.0 ml. of water and 2.0 ml. of 0.75 N KOH to 16.0 ml. of zinc sulfate reagent (No. 13a) in a Pyrex centrifuge tube, mix. Centrifuge, wash, and continue as directed in the procedure.

II. ALKALINE-ASH METHOD

Reagents

Reagents Nos. 1, 3, 4, 9, 10, 11, 13a and b, and 14 described under the Distillation Method are used in this method also. Omit sodium chloride from reagent No. 6.

15. 2 N Hydrochloric acid. Dilute 165 ml. concentrated HCl (reagent grade) to 1 l. with redistilled water.

16. 7 N Sulfuric acid. Dilute 200 ml. of concentrated H_2SO_4 (reagent grade) to 1 l. with redistilled water.

17. Hydrochloric acid and sulfuric acid mixture. Mix together 2 parts of 2 N HCl, 2 parts of 7 N H_2SO_4, and 3 parts of redistilled water. Use 7.0 ml. of this mixture to dissolve ash.

18. 4 N Potassium carbonate (reagent grade). Dissolve 276.0 g. of C.P. anhydrous K_2CO_3 in redistilled water and dilute to 1 l.

Testing of Additional Reagents of Alkaline Ash Method

1. 4 N Potassium carbonate and hydrochloric acid-sulfuric acid mixture. Mix 1.0 ml. of 4 N potassium carbonate, with 7.0 ml. of acid mixture (which should be approximately 2 N) and continue with 5.0 ml. as described in *Color Development* under the Distillation Method.

Procedure

PROTEIN PRECIPITATION

Add 1.00 ml. of serum (slightly hemolyzed is satisfactory) to 8.0 ml. of $ZnSO_4$ solution in a 16 x 150-mm. Pyrex test tube. (Pyrex test tubes other than 16 x 150 mm. may be used if good recoveries are obtained by their use.) Mix with a stirring rod and then add 1.00 ml. of 0.75 N potassium hydroxide. Mix thoroughly and let stand for 15 minutes. Centrifuge for 10 minutes at 2000 r.p.m. Pour off the supernatant. Wash once with 10 ml. of water (iodine-free) by suspending the precipitate and then centrifuging for 10 minutes at 2000 r.p.m. Add the washing to the first supernatant for the total iodine determination, if desired. Test the first supernatant as directed under the Distillation Method in PRECIPITATION OF PROTEIN AND WASHING OF PRECIPITATE.

DRYING AND ASHING

Mix 1.00 ml. 4 N potassium carbonate with the washed zinc protein precipitate, keep all of the carbonate mixture within 1 inch of bottom of the tube. Deposition of the carbonate as a layer on top of the precipitate without stirring has been suggested (16). Dry overnight (or until completely dry) in oven at 90–95°C. The time of drying may be shortened by using a forced-draft oven with an air filter. Incinerate the dried residue in a muffle furnace at 600° C. ± 25° for 2½ hours. (Some furnaces, because of their particular characteristics, may require additional incineration time.) The ash may or may not contain small amounts of carbon which can be separated by centrifugation after the addition of the acid mixture. Pyrex test tubes will usually last for three incinerations.

DISSOLVING IODIDE FROM ASH

Add 7.0 ml. of acid mixture to each ashing tube. Add the first 2.0 ml. slowly; twirl to hasten solution of the ash. When effervescence stops, mix thoroughly by twirling. This acid mixture should be approximately 2.0 N. Centrifuge until the supernatant is clear and the precipitate is well packed.

PHOTOMETRIC MEASUREMENT

Continue with 5.0 ml. of the ash solution as described for 5.0 ml. of distillate in the Distillation Method.

Calculation

Continue as described under the Distillation Method.

$$\text{Micrograms of iodine/100 ml.} = \frac{100}{1} \times \frac{7}{5} \times \text{conc. std}$$

$$\times \frac{\%T \text{ unknown } - \%T \text{ of blank}}{\%T \text{ of standard } - \%T \text{ of blank}}$$

(This calculation applies only to straight-line curve graph data.)

Discussion

Standardization and rates of reaction curves obtained with ceric hydroxide reagent under different conditions are presented in Figs. 2–7 in order to indicate the kind of data obtained with this reagent. The measurements were made with a No. 14 Coleman spectrophotometer. In the distillation procedure, a good spread in trans-

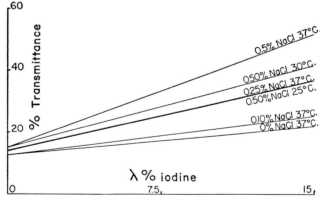

FIG. 2. The effect of different concentrations of sodium chloride in the 0.06 N arsenious acid reagent when it was used with 0.0114 N ceric reagent in the measurement of iodine in distillates from acid digestion of reagent blanks incubated 30 minutes at 25°, 30°, and 37°C.

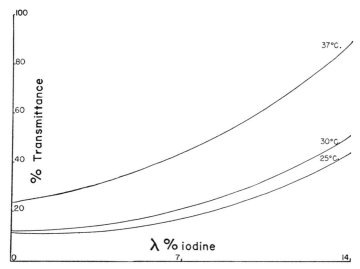

FIG. 3. The per cent transmittance obtained with $0.0500\,N$ ceric reagent in the measurement of iodine in the alkaline-ash residue of reagent blanks to which known amounts of iodine were added, when incubated 30 minutes at 25°, 30°, and 37°C.

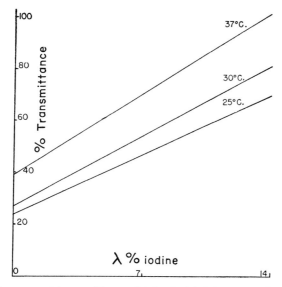

FIG. 4. The per cent transmittance obtained with $0.0114\,N$ ceric reagent in the measurement of iodine in the alkaline-ash residue of reagent blanks to which known amounts of iodine were added, when incubated 30 minutes at 25°, 30°, and 37°C.

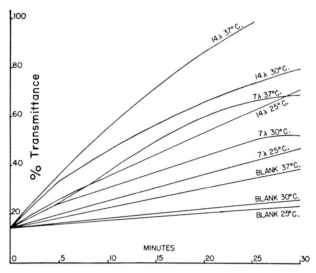

Fig. 5. The rate of reduction at different temperatures of 0.0114 N ceric reagent by iodine standards equivalent to 14 and 7 μg./100 ml. of hypothetical serum in the alkaline-ash method.

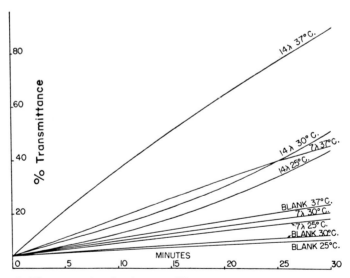

Fig. 6. The rate of reduction at different temperatures of 0.0500 N ceric reagent by iodine standards equivalent to 14 and 7 μg./100 ml. of hypothetical serum in the alkaline-ash method.

mittance readings and a straight-line standardization curve on a rectilinear coordinate graph is obtained at all temperatures by the addition of 0.1–0.5% sodium chloride to 0.0114 N ceric reagent, as shown in Fig. 2. However, sodium chloride cannot be used with 0.0500 N ceric hydroxide, as turbidity is produced. In the alkaline-ash method, 0.0500 N ceric hydroxide does not give a straight-line standardization curve at 25°, 30° and 37°C. when plotted on co-ordinate graph paper, as shown in Fig. 3. When plotted on a semilog graph at 37°C. the curve is linear. However, 0.0114 N ceric reagent gives a straight-line standardization curve at 25°, 30° and 37°C. on a rectilinear coordinate graph, as shown in Fig. 4. The rate of reduction of 0.0114 N ceric reagent in the alkaline-ash method is linear in blanks and in the 7-μg. standard at 25°C. but is not linear in other concentrations at higher temperatures (Fig. 5). The rate of reduction of 0.0500 N ceric reagent in the alkaline-ash method is linear in 7-μg. standards at 25° and 30°C., but other standards at higher temperatures were not linear (Fig. 6). The rate of reduction of 0.0114 N ceric reagent in the distillation method in the presence of added sodium chloride is linear in all blanks and standards at all temperatures (Fig. 7).

At the present time, reagent-grade ceric hydroxide is obtainable only from G. Frederick Smith Chemical Company. "The $Ce(OH)_4$ is made by a process that insures the absence of any impurities and is more costly than $(NH_4)_2Ce(SO_4)_3 \cdot 2 H_2O$ or $H_2Ce(SO_4)_3 \cdot H_2SO_4$

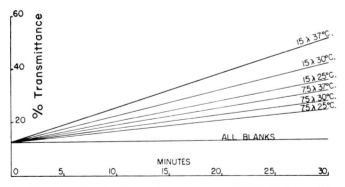

FIG. 7. The rate of reduction in the presence of 0.5% NaCl at different temperatures of 0.0114 N ceric reagent by iodine standards equivalent to 15 and 7.5 μg./100 ml. of hypothetical serum in the distillation method.

162   STANDARD METHODS OF CLINICAL CHEMISTRY II

or other salt or acid. The formula $Ce(OH)_4$ is near or exact enough to give close agreement of pre-meditated strength from a given weighed amount of sample reagent." (17). If difficulty is experienced in getting $Ce(OH)_4$ into solution, the following technique may be used: Grind the dry reagent to a fine powder in a mortar and treat with hot sulfuric acid [1 volume concentrated H_2SO_4 (95%) + 3 volumes of distilled water] and adjust the sulfuric acid to 1.8 N when dilution is made to the desired volume.

NOTE: The ceric ammonium sulfate which is made by the G. Frederick Smith Company is now generally considered to be of suitable purity and is more widely used in the measurement of PBI.

Precautions

In order to measure iodine accurately with a method sensitive to 0.005 μg. or 1 part of iodine to twenty billion, extreme caution and rigid control must be exercised in the selection and preparation of reagents and in the conditions specified in the standard method. All new stocks of reagents should be carefully tested before use on unknown specimens, regardless of source or manufacture.

All glassware must be carefully cleaned with chromic-sulfuric acid and, after rinsing, washed with dilute KOH before a final distilled water wash. This glassware should be kept free of dust and chemical fumes.

The digestion, distillation, and ashing apparatus should be isolated, during both use and storage, from possible contaminants, chemical fumes, dust, etc. All metal salts which give colored solutions will interfere with the reduction of ceric sulfate. Cyanides, and mercuric, silver, and manganese salts will also interfere.

Range of Normal Values

The generally accepted normal range for serum PBI is 4–8 μg./ 100 ml. It has been our experience that over 90% of all of our PBI determinations (over 12,000) on hospital patients range from 5 to 7 μg./100 ml. of serum. Published data (18–21) on the normal range of serum PBI has been reported as 3.4–8.0 μg./100 ml.

REFERENCES

1. Kolthoff, I. M., and Sandell, E. B., Chronometric catalytic method for the determination of micro quantities of iodine. *J. Am. Chem. Soc.* **56,** 1426 (1934).

2. Chaney, A. L., Improvements in determination of iodine in blood. *Ind. Eng. Chem., Anal. Ed.* **12**, 179–181 (1940).
3. Barker, S. B., and Humphrey, M. J., Clinical determination of protein-bound iodine in plasma. *J. Clin. Endocrinol.* **10**, 1136–1141 (1950).
4. Moran, J. J., Factors affecting the determination of protein-bound iodine in serum. *Anal. Chem.* **24**, 378–384 (1952).
5. Van Zyl, A., A critical assessment of the distillation technique for the estimation of protein-bound iodine with suggested improvements. *S. African J. Med. Sci.* **18**, 61–78 (1953).
6. Talbot, N. B., Butler, A. M., Saltzman, A. H., and Rodriguez, P. M., The colorimetric estimation of protein-bound serum iodine. *J. Biol. Chem.* **153**, 479–488 (1944).
7. Connor, A. C., Swenson, R. E., Park, C. W., Gangloff, E. C., Liebermann, R., and Curtis, G. M., The determination of blood iodine; A useful method for the clinical laboratory. *Surgery* **25**, 510–517 (1949).
8. Sobel, H., and Sapsin, S., Modified procedure for determination of protein-bound iodine in serum. *Anal. Chem.* **24**, 1829–1831 (1952).
9. Zak, B., Willard, H. H., Myers, G. B., and Boyle, A. J., Chloric acid method for determination of protein-bound iodine. *Anal. Chem.* **24**, 1345–1348 (1952).
10. O'Neal, L. W., and Simms, E. S., Determination of protein-bound iodine in plasma or serum. *Am. J. Clin. Pathol.* **23**, 493–505 (1953).
11. Zieve, L., Dahle, M., and Schultz, A. L., Comparison of incineration and chloric acid methods for determination of chemical protein-bound iodine. *J. Lab. Clin. Med.* **44**, 374–377 (1954).
12. Meyer, K. R., Dickerman, R. C., White, E. G., and Zak, B., Study of inhibition of the ceric-arsenite reaction and application to analysis of protein-bound iodine. *Am. J. Clin. Pathol.* **25**, 1160–1170 (1955).
13. Grossman, A. and Grossman, G. F., Protein-bound iodine by alkaline incineration and a method for producing a stable cerate color. *J. Clin. Endocrinol. & Metabolism* **25**, 354–361 (1955).
14. Chaney, A. L., Review of techniques for protein-bound iodine. *Clin. Chem.* **5**, 6–7 (1953).
15. Hyde, L., and Hyde B., Effect of retained bronchial lipiodol on blood iodine. *J. Lab. Clin. Med.* **34**, 1516–1519 (1949).
16. Skause, B. and Hendenskog, I., The determination of serum protein-bound iodine by alkali incineration. *Scand. J. Clin. & Lab. Invest.* **7**, 291–297 (1955).
17. Smith, G. F., Univ. Illinois. Personal communication. (1954)
18. Brown, H. Reingold, A. M., and Samson, M., The determination of protein-bound iodine by dry ashing. *J. Clin. Endrocrinol. & Metabolism* **13**, 444–450 (1953).
19. Kydd, D. M., Man, E. B., and Peters, J. P., Concentration of precipitable iodine in the serum. *J. Clin. Invest.* **29**, 1033–1040 (1950).
20. Barker, S. B., Humphrey, M. J., and Soley, M. H., The clinical determination of protein-bound iodine. *J. Clin. Invest.* **30**, 55–62 (1951).
21. Blackburn, C. M., and Power, M. H., Diagnostic accuracy of serum PBI

determinations in thyroid disease. *J. Clin. Endocrinol. and Metabolism* **15,** 1379–1392 (1955).

ADDITIONAL PERTINENT REFERENCES

Thompson, H. L., Klugerman, M. R., and Truemper, J., A method for PBI: The kinetics and the use of controls in the ashing technique. *J. Lab. Clin. Med.* **47,** 149–163 (1956).

Fischl, J., Determination of protein-bound iodine in micro-amounts of serum or plasma. *Clin. Chim. Acta* **1,** 462–469 (1956).

Sanz, M. C., Brechbühler, T., and Green, I. J., The ultramicro-determination of total and protein-bound iodine. *Clin. Chim. Acta* **1,** 570–576 (1956).

SODIUM AND POTASSIUM BY FLAME PHOTOMETRY

Submitted by: PAULINE M. HALD, Department of Internal Medicine, see MS 513
Yale University School of Medicine, New Haven, Connecticut
W. BURKETT MASON, Department of Biochemistry and Atomic
Energy Project, School of Medicine and Dentistry, University of
Rochester, Rochester, New York

Introduction and Principles

In recent years flame photometry has largely replaced chemical methods for determinations of sodium and potassium. Flame photometric methods have also been proposed for determinations of calcium (1–12) and of magnesium (4, 10, 12) but are not yet well enough established to be acceptable for general use in the clinical laboratory.

The basic principles of flame photometry are simple and generally well known. A solution of the element to be determined is sprayed into a flame. The flame temperature is sufficient to excite some of the resulting atoms to an electronic state which is above the ground state, and these excited atoms return to the ground state by emitting radiation of a characteristic wavelength. When all the variables are properly controlled, the intensity of the characteristic radiation is proportional to the concentration of the element. In practice, it is found that only a few elements are easily excited by flame techniques and that their characteristic radiations have quite different wavelengths.

The essential components of a flame photometer are thus a flame, a means of spraying (atomizing) the solution into the flame, an optical device for isolating the characteristic radiation, and a photometer for measuring the intensity of this radiation. Many satisfactory flame photometers are commercially available. They fall into two general classes: *direct-intensity instruments* and *internal-standard instruments*. It is not the purpose of this chapter to provide detailed descriptions of these instruments, for such descriptions

165

are readily obtained from the various manufacturers (13). Neither is there any intent to debate the relative merits of individual instruments. Technical aspects of flame photometer design have recently been reviewed by Gardiner (14) and by Margoshes and Vallee (15).

The best practical guides in choosing a flame photometer are the recommendations of experienced analysts in accredited medical laboratories who have firsthand knowledge of the instrument's performance. The distributor of any instrument should be able and willing to supply such references. Direct communication should be established with the person who has been responsible for the performance of the instrument and, if possible, it should be seen in operation under circumstances similar to those for which it is to be used. Advertising that makes broad claims of performance is often quite misleading and may be based upon experiences with simple inorganic solutions rather than analysis of complex biological materials.

The selection of a flame photometer should be made with respect to the particular needs of the laboratory. No one instrument is ideal for all applications, and features which are advantageous in one instance may be decidedly inferior at other times. For the clinical laboratory, where large numbers of analyses for serum sodium and potassium are expected and reliability is a prime concern, the most satisfactory arrangement probably will be to make the sodium and potassium determinations with an instrument having a relatively low-temperature flame and a photometer employing one or more barrier-layer photocells that are coupled directly to a sensitive galvanometer. Such an instrument is of simplest design, is least expensive, requires little maintenance, and employs standard solutions that are relatively simple to prepare and maintain. Accuracy for sodium and potassium is satisfactory for both clinical and investigative work. If it is also desired to do calcium and magnesium by flame photometry, it will probably be more satisfactory and more economical in the long run to have a separate, more complex instrument for determining these elements.

General Considerations

Plans for installation and operation of a flame photometer should provide for several weeks of intensive work with the instrument. This period of orientation is necessary for the person who will ultimately be responsible for its operation and maintenance to become familiar with the variable factors and their proper adjustment.

Location of the instrument should receive as much consideration as that of a fine analytical balance. Thermal conditions in the room should be stable. Intermittently operating radiators and drafts should be avoided. If possible the flame photometer should be placed in a hood or be equipped with a flue that will carry off the hot vapors and help in stabilizing room temperature. Incinerated material that settles about the room is a continual source of contamination, and from it glassware, solutions, and pipets may become sources of error. It is advisable to dust the room, the instrument, reagent bottles, desks, counters, and other nearby areas daily; this is done preferably with a damp cloth. The flues above the burners and the walls of the room should be vacuum cleaned from time to time, especially in those laboratories where large volumes of work are done.

In practice, the *atomizer* is probably the most critical component of the flame photometer. Unless the solution is carried to the flame in a steady and uniform manner, the emission of light will be unstable and the resulting fluctuations will be reflected in the recorded readings. Many types of atomizers and atomizing chambers have been designed. They are well summarized by Gardiner (14).

For the analysis of biologic materials, particularly serum and other protein-containing fluids, the atomizer must have certain characteristics beyond those essential for work with simple salt solutions. In particular, the atomizer must be rapidly self-cleaning, otherwise introduction of a few samples will lead to difficulties from irregular flow. Tiny particles of protein, precipitated material, dust, or air bubbles may interrupt the flow of solution and thus give erratic results. It is imperative that the atomizer be easily removable to facilitate regular soaking in a suitable solvent. It is also advantageous if more frequent cleaning can be made with the atomizer in place, either by rinsing with solvent or by probing with a stylet, to dislodge particles or air bubbles. A small orifice permits economy of material, but is likely to lead to intolerable plugging and consequent poor operation.

Successful operation of a flame photometer is impossible without good control of the *flame*. Pressures of the various gases must be well regulated and the gases themselves must be free of impurities. The importance of constant and satisfactory flame conditions cannot be overemphasized. The appearance of the flame is an important guide to its performance, and the operator should become familiar with the appearance of the flame at optimum performance. Whenever the response of the instrument is not entirely satisfactory, the flame should be observed. If its appearance is not ideal, steps should be taken to correct whatever irregularity may be responsible. In those instruments with a chimney that separates the flame from the photometer unit, it is well to inspect the chimney frequently, wash it whenever necessary, and replace it if the window portions become etched or cloudy.

All gases that are employed should be filtered. The instrument manufacturer sometimes supplies a built-in filter or recommends a commercial type that is particularly well suited to his product. It is sometimes rewarding to interpose additional filters in the line, since impurities may find their way into the flame even though a filter is in use. It is always possible to set up a simple filter by passing the gas through a carboy of water. It is important to inspect filters at regular intervals and to replace them whenever it appears necessary. When a tank of oxygen, propane, or other fuel is replaced, it is advisable to open the valve momentarily to expel any small particles that might subsequently clog the regulator if forced out after the tank has been connected to the flame photometer. It is also well to wipe out the valve orifice before connecting the tank.

Some fire insurance companies require that cylinders of certain fuel gases be located outside the building and permit only specified kinds of piping for bringing these gases into the laboratory. Failure to consider such regulations, as well as various provisions of local building codes, etc., may lead to unnecessary inconvenience in installing a flame photometer or in converting to a newer model.

The *accuracy* of the instrument's performance must be known before it can be considered acceptable for regular use. For serum, sodium determinations should be accurate to $\pm 1.5\%$ and for potassium, to $\pm 5\%$. Greater accuracy does not add appreciably to the value of these measurements. Less accurate results may be obtained with more complex materials, e.g. food, feces, tissues. When good blending is achieved, however, the chief deterrent to accuracy is the effect of interfering ions.

Neither *precision* nor *stability of performance* should be confused with *accuracy*. Precision refers to the reproducility of results, without regard to their correctness. Many photometers, especially those using the internal-standard principle, are capable of a precision corresponding to uncertainties of less than 1% for both sodium and potassium measurements. The over-all accuracy of the determination can approach, but not exceed this figure. Some instruments appear to exhibit very stable operation, and the readings are exactly reproducible. This effect, however, may be the result of excessive damping of the instrument. Frequently a less stable (wider fluctuating needle) but more sensitive instrument will produce more accurate results.

Accuracy can be evaluated by two procedures: (a) assay of solutions containing approximate amounts of the various inorganic salts present in serum (or the material under examination) and accurately known amounts of the elements to be determined; and (b) replicate determinations on different dilutions of the same biologic sample. Both of these methods have been incorporated into the procedures described in this chapter.

Blank determinations are important and are as necessary to detect contaminating or interfering substances in flame photometry as in more conventional chemical procedures. Whenever materials are ashed or extracted, all the solutions used are possible sources of contamination. It is especially important to guard against contamination from distilled water. This source of error sometimes arises quite unexpectedly, and it is good practice to test the conductivity of the distilled water daily. Alternatively, a supply of distilled water can be collected in an appropriate container and used for diluting both the working standards and the samples to be analyzed. New working standards must be prepared whenever the distilled water is replenished. Filter paper must be tested for the elements being determined, and it may be necessary to wash and dry the filter paper before it is used.

Control samples, consisting of pooled sera or of reconstituted serum, are used in some laboratories. If these are available, daily analysis should be made by all the flame photometric procedures in use.

Collecting, Storing, and Preparing Samples

Serum for the determination of electrolytes should ordinarily be obtained when the patient is in a fasting state, i.e. preferably in the morning before breakfast. Venous blood should be drawn without stasis and by preventing loss of CO_2 (16–19). A dry syringe which has been rinsed with mineral oil should be used, and the blood should be collected with minimal evacuation of the syringe. The needle is removed and the tip of the syringe is placed against the side of the collecting tube beneath the surface of the mineral oil that is already present. The blood is then gently expelled without introducing air bubbles. There should be no hemolysis, and the serum should be separated at once to prevent transfer of electrolytes between cells and serum (16, 20, 21). No further change in sodium or potassium is expected after the serum has been separated. It is advisable to deep-freeze samples which are to remain any length of time before analysis, as growth of molds and precipitation of protein may make accurate measurements difficult if the sample is allowed to remain for many days at room temperature, or even in the refrigerator. Samples which have been frozen should be allowed to become fluid at room temperature. The container is then inverted at least twenty times to insure homogeneous composition.

Urine should best be analyzed soon after collection. If it is stored in a refrigerator, the precipitation of various constituents, especially phosphates, may make it necessary to filter the sample or dissolve the precipitate by chemical means in order to remove particles before introducing the solution into the flame photometer. Any material removed by filtration should be tested for the elements being determined before being discarded. Deep freezing is advisable for extended periods of storage.

Food, feces, and tissues require special treatment that varies with the particular program in progress. All of these materials must be homogenized to insure representativeness of the samples taken for analysis. For large volumes of materials, especially food and feces, a blender that accommodates various sized receptacles is convenient; the *Osterizer*,[1] which handles jars up to a gallon in size, has been found very satisfactory. For tissues, many devices have been designed to suit the needs of individual analysts.

The samples taken for analysis must be ashed or extracted to convert the electrolytes into a form suitable for introduction into the flame photometer. Dry ashing in a muffle furnace (22) is time-consuming and requires special apparatus. With some materials (e.g. bones), however, it is the procedure of choice, and, once the routine has been established, dry ashing is both simple and satisfactory. It is often more convenient and expedient to extract the material with ni-

---

[1] John Oster Manufacturing Company, Racine, Wisconsin.

tric acid. The extract may be stored until needed. If the samples taken for analysis are not processed immediately, they had best be kept frozen.

PROCEDURE FOR NITRIC ACID EXTRACTION (23)

Transfer an aliquot of homogenized mixture equivalent to approximately 5 g. of original material to a 250-ml. Erlenmeyer flask and mix with 3 ml. of concentrated nitric acid. After heating momentarily to the boiling point and then allowing to cool, transfer the contents quantitatively to a 100-ml. volumetric flask, dilute to volume with water, mix thoroughly, and filter through dry paper which is known to be free from the elements being determined. Alternatively, a sintered glass disk may be used for preparing the filtrates. For the analysis, transfer a suitable aliquot (usually 10 ml.) of filtrate to a volumetric flask of appropriate size (usually 100-ml.) and dilute to volume with water after first adding any solutions that are required for the flame photometer being used. With some instruments, in order to reduce corrosion of atomizer and burner components, it may be advisable to neutralize part of the acid with ammonium hydroxide before making the solution to volume. Appropriate steps must then be taken to correct for any contaminant introduced with the ammonium hydroxide.

Dilution of materials is necessary to bring the concentrations into ranges that are optimal for accurate analysis. The sodium and potassium concentrations of serum are limited to a relatively small range, and it is not difficult to determine the exact dilution required. The concentrations in urine, food, and feces, however, vary widely, and the appropriate dilutions must be determined by trial. It is usually advisable to start with a rather small dilution. A second dilution can then be made if the first reading exceeds the concentration range of the standards.

In general it is advisable to dilute to a concentration that lies within the lowest range of optimal performance of the instrument. Several advantages are to be gained: (a) Minimal amounts of material are required. (b) The effects of viscosity and interference from foreign ions are reduced. (c) The calibration curve tends to become more nearly linear at lower concentrations.

The chief disadvantage of working at low concentrations is the increased danger of contamination.

Whenever possible, it is good practice to read a given unknown at two different dilutions. In this way small discrepancies in the concentration of the standard solutions, the presence of contaminants, or errors in dilution become apparent. Replicate measurements with different concentrations constitute the most rigid check on various sources of error. In determining serum sodium, dilutions of 1:250 and 1:500 will yield results that agree within 1 to 1½% when conditions are optimal.

Reagents

Specific reagents are usually recommended by the manufacturer for each model of flame photometer, and these solutions are described in the brochure which accompanies the instrument. All of the instruments require concentrated stock solutions, and from these more dilute working solutions are prepared. Certain basic principles apply to the preparation of all these solutions.

Good analytical techniques should be employed, for the ultimate accuracy of the determinations can not exceed the accuracy of the standard solutions. Volumetric apparatus should be accurately calibrated, and attention should be paid to making all solutions uniform. A safe rule is to invert and shake each container twenty times after the solution has been diluted to the mark. It is important that salts which are used be of highest grade and that the amounts taken for preparation of standards be carefully dried and accurately weighed. The distilled water should be free of any contaminant which will interfere with the determination. If the distilled water at hand contains appreciable amounts of the element being determined, it should be purified by distillation or by use of ion-exchange columns. Conductivity measurements are easy to make and provide a convenient check on the ionic purity of the distilled water. Bottles for storing both standard solutions and samples should be of Pyrex, No-Sol-Vit glass, polyethylene, or other material which will not introduce contamination.

STOCK SOLUTIONS

The following stock solutions have been found satisfactory for many different instruments.

1. Sodium chloride (50 meq./l.). Transfer 2.922 g. of NaCl that has been dried to constant weight quantitatively to a 1-l. volumetric flask, dilute to volume with water, and make the solution uniform.

2. Potassium chloride I (50 meq./l.). Transfer 3.727 g. of KCl that has been dried to constant weight quantitatively to a 1-l. volumetric flask, dilute to volume with water, and make the solution uniform.

3. Potassium chloride II (10 meq./l.). Transfer 0.7455 g. of KCl that has been dried to constant weight quantitatively to a 1-l. volumetric flask, dilute to volume with water, and make the solution uniform.

4. Lithium nitrate (10 g. Li/l.). Transfer 99.35 g. of LiNO₃ that has been dried to constant weight quantitatively to a 1-l. volumetric flask, dilute to volume with water, and make the solution uniform.

Lithium nitrate is used only in the internal-standard procedure, and slight variations in Li concentration are acceptable. It is imperative, however, that exactly the same Li concentration be included in working standards and unknowns. For this reason it is convenient to prepare the lithium nitrate solution as accurately as the sodium and potassium stock solutions so that newly prepared solutions will be of the same concentration as those already in use.

5. "Artificial serum" (138.1 meq. Na/l. and 4.10 meq. K/l.).

NaCl	8.072 g.
KCl	0.2091 g.
KH₂PO₄	0.1768 g.
CaCl₂.2H₂O	0.37 g.
MgSO₄.7H₂O	0.25 g.

Dry and weigh accurately only the sodium and potassium salts, as "artificial serum" is not intended as a standard for either calcium or magnesium. Transfer the salts quantitatively to a 1-l. volumetric flask, dilute to volume with water, and make the solution uniform.

"Artificial serum" contains the inorganic salts present in serum in concentrations similar to those occurring normally. It is used to test the performance of the flame photometer and is analyzed in exactly the same manner as serum. The lack of protein detracts in no way from the useful function of this solution as serum is so highly diluted before analysis.

Direct-Intensity Instruments

Principle of operation: In direct-intensity instruments, a solution containing the element to be determined is atomized at constant rate into a well-controlled flame, and the intensity of the radiation emitted at the characteristic wavelength for that element is measured. Since the relationship between radiation intensity and concentration is not linear, it is desirable to construct a calibration curve with closely spaced points. In the directions which follow, it is assumed that the indicating unit of the photometer is a galvanometer.

Determination of Serum Sodium Using a Direct-Intensity Instrument

Working Standards

Prepare a series of solutions containing 0.25, 0.50, 0.75 and 1.00 meq. of sodium per liter by measuring exactly 5.0-, 10.0-, 15.0-, and 20.0-ml. portions of stock sodium chloride (50 meq./l.) into 1-l. volumetric flasks and diluting to volume with water. Set aside some of the *same water* for use in making the zero setting on the flame photometer and for diluting both the "artificial serum" and the serum samples. Alternatively, the purity of the distilled water may be proved by conductivity measurements or by determining the sodium content by flame photometry.

Dilution of "Artificial Serum" 1:250 and 1:500

Measure 1.00-ml. portions of "artificial serum" into 250- and 500-ml. volumetric flasks and dilute to volume with appropriate water.

Dilution of Serum 1:250 and 1:500

Dilute 0.200-ml. portions of serum to 50 and 100 ml., using volumetric flasks and appropriate distilled water.

Method of analysis

With the flame burning properly and the instrument set to the sodium wavelength, the galvanometer is set at zero while distilled water is being atomized into the flame. Working standard 1.00 meq./l. is then substituted for the water, and the necessary photometer adjustments are made to give full deflection of 100 scale units on the galvanometer. Distilled water is again introduced and the zero setting checked. If necessary, the zero is reset; the 100 mark may then also need to be reset. These two readings are checked several times. Unless both readings remain constant on successive occasions, the instrument is not operating satisfactorily and must be adjusted until stable conditions are attained. When these points are satisfactory, the intermediate standards 0.25, 0.50, and 0.75 meq./l. are introduced successively, and the galvanometer readings recorded. The entire series, including the zero and the 1.00 meq./l. solutions should be reread several times so that any gross variation which may occur later will be recognized.

A plot of galvanometer readings against concentration of the working standards gives a calibration curve for the serum sodium determination. The chief use of this calibration curve is to judge the operation of the instrument and to insure that the various working standards have been properly prepared. If the values for the entire series of standards are perfectly reproducible, it is possible to use this calibration curve directly for determining the unknown concentrations. It is advisable, however, to check the instrument's performance by including several readings on the diluted "artificial serum" and to check the calibration points during and following the unknown measurements.

In practice, it is usually found that small variations in readings for the standards occur so frequently that the procedure of "bracketing" is advisable. In this procedure a standard having a concentration below that of the unknown is read, then the unknown, and finally a standard having a higher concentration, or the order may be reversed. Sandwiching the unknown between the two standards has the advantage of minimizing the effects due to instrument drift. The unknown concentration is determined from the standard concentrations by interpolation of the readings. This is conveniently done graphically by a procedure to be explained below. Since the interpolation is linear while the calibration curve is usually slightly bowed, close agreement between readings for standards and unknown is necessary for the most accurate results. For this reason some instrument designers have recommended a large series of standard solutions so that each unknown may be read in close proximity to a standard. This is very desirable, especially when small changes are under consideration, but the advantages must be weighed against the added work required to maintain the larger number of solutions.

It is important to emphasize that "bracketing" provides a separate calibration for each unknown and is therefore the procedure of choice whenever greatest accuracy is desired.

Method of "bracketing" applied to "artificial serum"

It is known that the readings should fall between 0.25 and 0.50 for the 1:500 dilution and between 0.50 and 0.75 for the 1:250 dilution. The zero point is checked or reset, the 0.25 standard is read, then the sample that was diluted 1:500, followed by the 0.50 standard. The 1:250 dilution is next read, followed by the 0.75 standard. The readings are taken to the nearest $\frac{1}{4}$ division and recorded in the following manner:

$$[S_1 \quad C_1 \quad S_2 \quad C_2] \quad U$$

S_1 is the reading for the first standard and C_1 is the corresponding concentration; S_2 is the reading for the second standard and C_2 is the corresponding concentration; U is the unknown reading. The two points (S_1, C_1) and (S_2, C_2) are plotted on large graph paper and connected by a straight line as in Fig. 1. Coordinate paper ruled ten squares to the inch is very satisfactory. The concentration C, corresponding to the reading of the unknown sample, is determined from

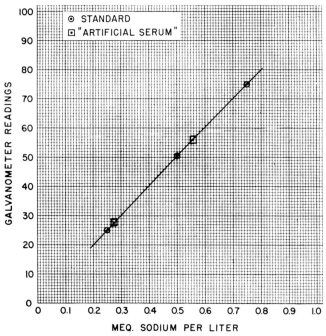

FIG. 1. Graphical method for determining unknown concentration from readings obtained by "bracket" procedure. Data are for sodium in "artificial serum" using a direct-intensity instrument.

this plot, and represents the concentration of sodium in the diluted solution. This value is multiplied by the dilution factor, D, to give the concentration in the original solution, i.e.:

$$C \times D = \text{Meq./l.}$$

Alternatively, the concentration in the original solution can be calculated as follows:

$$\left[C_1 + \frac{(U - S_1)(C_2 - C_1)}{(S_2 - S_1)} \right] \times D = \text{Meq./l.}$$

The symbols are as defined previously. It makes no difference whether C_1 or C_2 refers to the higher concentration so long as proper attention is paid to the mathematical signs which result.

The following data are typical of those obtained with "artificial serum."

Dilution	S_1	C_1 meq./l.	S_2	C_2 meq./l.	U	$C \times D$	meq./l.
1:500	25.00	0.25	50.50	0.50	27.50	0.275 × 500	138
1:250	50.50	0.50	75.00	0.75	56.00	0.556 × 250	139

The analysis of serum should not be undertaken until satisfactory results have been obtained for "artificial serum."

Method of "bracketing" applied to serum

Readings are obtained with diluted serum in the same manner as for "artificial serum." Examples for serums having various sodium concentrations are given in Table I.

TABLE I

SERUM SODIUM DETERMINATIONS—DIRECT-INTENSITY INSTRUMENT

Serum	Dilution	S_1	C_1 meq./l.	S_2	C_2 meq./l.	U	$C \times D$	meq./l.
A	1:250	25.00	0.25	49.50	0.50	48.50	0.490 × 250	123
	1:500	00.00	0.00	25.00	0.25	25.00	0.250 × 500	125
B	1:500	26.00	0.25	52.00	0.50	29.00	0.279 × 500	140
	1:250	52.25	0.50	76.25	0.75	58.00	0.560 × 250	140
C	1:250	76.25	0.75	51.00	0.50	63.25	0.622 × 250	156
	1:500	51.00	0.50	25.25	0.25	31.75	0.313 × 500	157
D	1:250	50.00	0.50	25.00	0.25	46.76	0.467 × 250	117
	1:500	25.00	0.25	00.00	0.00	23.50	0.235 × 500	118

Determination of Serum Potassium Using a Direct-Intensity Instrument

Potassium is present in low concentration in serum, and dilutions of 1:12.5 and 1:25 are recommended. At these dilutions the addition of water results in the precipitation of protein, which is undesirable. This can be avoided by making the dilutions with 0.9% NaCl (physiological saline). The use of saline, however, introduces a small blank, and an appropriate correction must be made. Rather than read this small correction as an absolute value, readings are taken on "artificial serum" which has been diluted both with saline and with water. The difference in readings (i.e. the blank) is of the order of $\frac{1}{2}$ or $\frac{1}{4}$ of a scale division on the galvanometer and is subtracted from the readings obtained with serum before the unknown concentration is read from the calibration curve. The blank remains constant as long as a given preparation of saline is used for diluting the samples.

Working Standards

Prepare a series of solutions containing 0.125, 0.25, 0.50, 0.75, and 1.00 meq. of potassium per liter by measuring exactly 2.5-, 5.0-, 10.0-, 15.0-, and 20.0-ml. portions of stock potassium chloride I (50 meq./l.) into 1-l. volumetric flasks and diluting to volume with distilled water. Observe the same precautions regarding distilled water as in the sodium determination.

Saline

Dissolve 18 g. of NaCl in 2 l. of water. The quantities need not be measured accurately.

Dilution of "artificial serum" 1:12.5 and 1:25

Measure 2.00- and 1.00-ml. portions of "artificial serum" into 25-ml. volumetric flasks and dilute to volume with water.

Dilution of serum 1:12.5 and 1:25

Measure 2.00- and 1.00-ml. portions of serum into 25-ml. volumetric flasks and dilute to volume with saline. If serum is present in insufficient quantity, duplicate dilutions of 1:25 are made.

TABLE II

SERUM POTASSIUM DETERMINATIONS—DIRECT-INTENSITY INSTRUMENT

Serum	Dilution	S_1	C_1 (meq./l.)	S_2	C_2 (meq./l.)	U	U Cor	$C \times D$	Meq./l.
"Artifi-	1:12.5	22.75	0.25	44.50	0.50	30.00	—	0.333×12.5	4.2
cial"	1:25	11.25	0.125	23.25	0.25	15.25	—	0.167×25	4.2
H	1:25	12.00	0.125	0.00	0.00	11.50	11.00	0.115×25	2.9
	1:25	0.00	0.00	12.00	0.125	11.00	10.50	0.110×25	2.8
J	1:25	11.25	0.125	22.50	0.25	14.50	14.00	0.156×25	3.9
	1:25	22.50	0.25	11.25	0.125	15.00	14.50	0.161×25	4.0
W	1:25	24.25	0.25	12.00	0.125	23.00	22.50	0.232×25	5.8
	1:25	10.75	0.125	22.50	0.25	20.75	20.25	0.226×25	5.7

Method of analysis

Construct a calibration curve covering the range 0 to 1 meq. potassium per liter in a manner similar to that described for sodium. When the points seem satisfactory, check the instrument's operation with the "artificial serum" and then proceed to analysis of the serum samples. The "bracket" procedure should be used. Typical data for "artificial serum" and serums having various potassium concentrations are tabulated in Table II. The blank correction for the serum readings is ½ scale division.

Internal-Standard Instruments

Principle of operation

The principle is similar to that of direct-intensity instruments, except that the intensity of the characteristic radiation for the element to be determined is compared to the intensity of the characteristic radiation for an "internal-standard" element which is present in known amount. Lithium is preferred as an internal standard for determination of sodium and potassium because it has excitation characteristics similar to those of these elements. Slight fluctuations in the discharge of the atomizer, or in the vigor of the flame, therefore affect the radiation intensity from both unknown and internal standard by approximately the same amount, and their intensity ratio is not appreciably altered. The effects of viscosity and interference by other ions are also minimized in the internal-standard method.

The ratio of radiation intensities from "internal standard" and unknown is usually determined by a null-point or balancing technique that employs a sensitive galvanometer. The internal-standard instruments are commonly designed in such a manner that increasing dial readings for the "balance point" indicate increasing amounts of the element to be determined, and this relationship will be assumed in the example which follows.

The proper amount of lithium to be added as internal standard depends on several factors and must be determined by experiment for a given application of a particular instrument. Usually the manufacturer will specify the optimal lithium concentrations for various uses. If it is desired to select a lithium concentration for some special situation, this should be done by experimentally determining the lithium concentration which gives the best compromise between calibration sensitivity and operational stability for standard solutions covering the desired range.

Since equal amounts of lithium solution are added to both standards and unknowns, it is not essential that the concentration of lithium be known exactly. In practice, however, it is advantageous to observe quantitative precautions so that successive batches of lithium solution may be used more or less interchangeably.

Determination of Serum Sodium and Potassium Using an Internal-Standard Instrument

In the example which follows, it is assumed that a lithium concentration of 100 mg./l. is satisfactory for sodium in the concentration range 0.00–0.75 meq./l. and for potassium in the concentration range 0.00–0.05 meq./l.

WORKING SOLUTIONS

Sodium standards

Prepare a series of solutions containing 0.125, 0.25, 0.50, and 0.75 meq. Na per liter and 100 mg. Li per liter by measuring exactly 2.50-, 5.00-, 10.0-, and 15.0-ml. portions of stock sodium chloride (50 meq./l.) into 1-l. volumetric flasks that contain 10.0 ml. of stock lithium nitrate (10 g. Li/l.). Dilute to volume with water and make the solutions uniform. Observe appropriate precautions regarding the water.

Potassium standards

Prepare a series of solutions containing 0.01, 0.02, 0.03, 0.04, and 0.05 meq. K per liter and 100 mg. Li per liter by measuring exactly 1.00-, 2.00-, 3.00-, 4.00-, and 5.00-ml. portions of stock potassium chloride II (10 meq./l.) into 1-l. volumetric flasks that contain 10.0 ml. of stock lithium nitrate (10 g. Li/l.). Dilute to volume with water and make the solutions uniform.

Zero solution

Dilute 10.0 ml. of stock lithium nitrate (10 g. Li/l.) to exactly 1 l. using the same water employed for preparing the working standards.

Dilute Lithium Solution (1 g. Li/l.)

Measure exactly 200 ml. of stock lithium nitrate (10 g. Li/l.) into a 2-l. volumetric flask. Dilute to volume with the same distilled water used to prepare the other working solutions and make the solution uniform.

Dilution of "Artificial Serum" 1 :500, 1 :250, and 1 :125

Measure 0.50, 1.00, and 2.00 ml. of "artificial serum" into 250-ml. volumetric flasks that contain 25.0 ml. of dilute lithium solution above and dilute to volume with appropriate water.

Dilution of Serum 1 :500, 1 :250, and 1 :125

Measure exactly 0.200-ml. portions of serum into 100-, 50-, and 25-ml. volumetric flasks containing respectively 10.0, 5.00, and 2.50 ml. of dilute lithium solution and dilute to volume with appropriate water.

METHOD OF ANALYSIS

The method of analysis is similar to that described for determination of serum sodium by the direct-intensity method, except that the zero position for the dial reading is set with the zero solution rather than with water. The calibration curve may be sufficiently reproducible to require only an occasional check against the stand-

TABLE III

SERUM SODIUM DETERMINATIONS—INTERNAL-STANDARD INSTRUMENT

Serum	Dilution	S_1	C_1 (meq./l.)	S_2	C_2 (meq./l.)	U	$C \times D$	Meq./l.
"Artificial"	1:500	156	0.25	296	0.50	173	0.280 × 500	140
	1:250	296	0.50	456	0.75	329	0.552 × 250	138
C	1:500	175	0.25	316	0.50	178	0.255 × 500	128
	1:250	316	0.50	481	0.75	325	0.514 × 250	129
D	1:500	164	0.25	314	0.50	187	0.288 × 500	144
	1:250	323	0.50	491	0.75	377	0.580 × 250	145
E	1:250	296	0.50	452	0.75	316	0.533 × 250	133
	1:500	165	0.25	294	0.50	172	0.263 × 500	132

TABLE IV

SERUM POTASSUM DETERMINATIONS—INTERNAL-STANDARD INSTRUMENT

Serum	Dilution	S_1	C_1 (meq./l.)	S_2	C_2 (meq./l.)	U	$C \times D$	Meq./l.
"Artificial"	1:250	47	0.010	106	0.020	85	0.0163 × 250	4.1
	1:125	157	0.030	202	0.040	173	0.0335 × 125	4.2
R	1:250	56	0.010	114	0.020	59	0.0105 × 250	2.6
	1:125	114	0.020	163	0.030	118	0.0208 × 125	2.6
M	1:250	77	0.020	—	—	77	0.0200 × 250	5.0
	1:125	112	0.030	153	0.040	151	0.0395 × 125	4.9
E	1:250	58	0.010	111	0.020	89	0.0159 × 250	4.0
	1:125	162	0.030	208	0.040	164	0.0303 × 125	3.8

ards; however, it is generally safer to use the "bracketing" procedure described for the direct-intensity method. Examples of typical data for serum sodium and potassium are given in Tables III and IV. Dilutions 1:500 and 1:250 were used for sodium measurements and dilutions 1:250 and 1:125, for potassium measurements. The calculations are as outlined for the sodium determination by the direct-intensity procedure.

Values in Healthy Persons

The concept of normal values for serum sodium and potassium requires some clarification. Various authors have included series of normal values in their publications, often without further comment.

Others refer to various figures as if the values were well defined and universally accepted. Tabulation of such data will show·variations of appreciable magnitude. In the 1954 International Biochemical Trial (24), in which the *same samples* were analyzed in over 100 laboratories throughout the world, the variations for both sodium and potassium exceeded acceptable limits of accuracy. This is a regrettable situation.

For the present, at least, it seem imperative that each laboratory compile its own series of "normal values" for use in evaluating serum samples that come to it. The group selected as subjects should be members of the laboratory, nursing, or medical staffs who are in good health, and not patients who are in the hospital for medical or surgical care. The blood should be collected, stored, and analyzed by the techniques in routine use. In such a series (25) compiled at Yale University, the maximum, minimum, and mean values for serum sodium in adults were 144, 131, and 138 meq./l. For serum potassium, the corresponding values were 4.6, 3.6, and 4.1 meq./l. These values are in good agreement with those given by Albritton (26).

Sodium and potassium values for about 500 foods and 150 public water samples have been determined by flame photometer methods. In reporting these values, Bills *et al.* (27) included comparisons between their method and the chemical techniques upon which most of the accepted values are based.

Sodium and potassium values for urine and feces depend upon intake and consequently are extremely variable.

Comment

An attempt has been made to present a general consideration of flame photometry as it is applied in the clinical laboratory. Emphasis has been placed upon various sources of error in the hope that others may profit by the experiences of the authors and their associates. This may give the impression that flame photometry is unwieldy. Nothing could be further from the truth, for once a sound procedure has been established and the routine well organized, great numbers of determinations are easily carried out with little difficulty.

ACKNOWLEDGMENTS

The authors are indebted to Dr. Augusta B. McCoord, Department of Pediatrics, School of Medicine and Dentistry, University of Rochester, for criticizing the manuscript and calling attention to several potential pitfalls that had been over-looked.

Parts of the chapter have been taken from a previous paper by the senior author (25) and are reproduced by permission of The Year Book Publishers, Inc., Chicago.

REFERENCES

1. Baker, R. W. R., The determination of calcium in serum by flame photometry. *Biochem. J.* **59**, 566–571 (1955).
2. Chen, P. S., Jr., and Toribara, T. Y., Determination of calcium in biological material by flame photometry. *Anal. Chem.* **25**, 1642–1644 (1953); Some errors in the determination of calcium in aged blood serum. *Anal. Chem.* **26**, 1967–1968 (1954).
3. Elert, B. T., A flame photometric method for calcium determination by direct serum dilution and comparison with the permanganate titration method. *Am. J. Med. Technol.* **21**, 297–303 (1955).
4. Kapuscinski, V., Moss, N., Zak, B., and Boyle, A. J., Quantitative determination of calcium and magnesium in human serum by flame spectrophotometry. *Am. J. Clin. Pathol.* **22**, 687–691 (1952).
5. Kingsley, G. R., and Schaffert, R. R., Direct microdetermination of sodium, potassium and calcium in a single biological specimen. *Anal. Chem.* **25**, 1738–1741 (1953); Micro flame photometric determination of sodium, potassium, and calcium in serum with organic solvents. *J. Biol. Chem.* **206**, 807–815 (1954).
6. MacIntyre, I., *Rec. trav. chim.* **74**, 498 (1955).
7. Masher, R. E., Intano, M., Boyle, A. J., Myers, G. B., and Isiri, L. T., The quantitative estimation of calcium in human plasma by flame spectrophoto-meter. *Am. J. Clin. Pathol.* **21**, 75–80 (1951).
8. Roth, C. F., and Sapirstein, L. A., A self-standardization method for flame spectrophotometric determinations of calcium in biological fluids. *Am. J. Clin. Pathol.* **25**, 1076–1089 (1955).
9. Severinghaus, J. W., and Ferrebee, J. W., Calcium determination by flame photometry, methods for serum, urine and other fluids. *J. Biol. Chem.* **187**, 1050–1051 (1954).
10. Vallee, B. L., Simultaneous determination of sodium, potassium, calcium, magnesium and strontium by a new multichannel flame spectrometer. *Nature* **174**, 1050–1051 (1954).
11. Winer, A. D., and Kuhns, D. M., Calcium determination by flame spectro-photometry. *Am. J. Clin. Pathol.* **23**, 1259–1262 (1953).
12. Zak, B., Mosher, R., and Boyle, A. J., A review of flame analysis in the clinical laboratory. *Am. J. Clin. Pathol.* **23**, 60–77 (1953).
13. Baird Associates, Inc., Cambridge, Massachusetts; Barclay Instruments Division, The Patent Button Company, Waterbury, Connecticut; Beckman Instruments, Inc., South Pasadena, California; Coleman Instruments, Inc., Maywood, Illinois; Instruments Division, The North American Philips Company, Inc., Mount Vernon, New York; Janke Aircraft Engine Test Equipment Company, Hackensack, New Jersey; The Perkin-Elmer Corporation, Norwalk, Connecticut; Process and Instruments Company, Brooklyn, New

York; Scientific Instruments Division, Fearless Camera Corporation, Los Angeles, California.

This listing is not intended to be exhaustive; it represents the instruments known to the authors.

14. Gardiner, K. W., Flame photometry. *In* "Physical Methods in Chemical Analysis" (W. G. Berl, ed.), Vol. 3, pp. 220–281. Academic Press, New York, 1956.

15. Margoshes, M., and Vallee, B. L., Flame photometry and spectrometry: Principle and applications. *In* "Methods of Biochemical Analysis" (D. Glick, ed.), Vol. 3. pp. 353–407 Interscience, New York, 1956.

16. Peters, J. P., and Van Slyke, D. D. "Quantitative Clinical Chemistry" Vol. 2, pp. 52–58. Williams & Wilkins, Baltimore, Maryland, 1932.

17. Kolmer, J. A., Spaulding, E. H., and Robinson, H. W., "Approved Laboratory Technic," 5th ed., p. 1003. Appleton–Century-Crofts, New York, 1951.

18. Goodale, R. H., "Clinical Interpretation of Laboratory Tests," p. 89. Davis, Philadelphia, Pennsylvania, 1955.

19. Miller, S. E., ed., "Textbook of Clinical Pathology," 5th ed., p. 375, Williams & Wilkins, Baltimore, Maryland, 1955.

20. Danowski, T. S., The transfer of potassium across the human blood cell membrane. *J. Biol. Chem.* **139,** 693–705 (1941).

21. Goodman, J. R., Vincent, J., and Rosen, I., Serum potassium changes in blood clots. *Am. J. Clin. Pathol.* **24,** 111–113 (1954).

22. Hald, P. M., The flame photometer for the measurement of sodium and potassium in biological fluids. *J. Biol. Chem.* **167,** 499–510 (1947).

23. Lowry, O. H., and Hastings, A. B., Histochemical changes associated with aging. I. Methods and calculations. *J. Biol. Chem.* **143,** 257–269 (1942).

24. Wooton, I. D. P., International biochemical trial, 1954. *Clin. Chem.* **2,** 296–301 (1956).

25. Hald, P. M., *In* Determinations with flame photometer. "Methods in Medical Research" (M. B. Visscher, ed.), Vol. 4. Year Book, Chicago, Illinois, 1951.

26. Albritton, E. C., ed., "Standard Values in Blood," Saunders, Philadelphia, Pennsylvania, 1952.

27. Bills, C. E., McDonald, F. G., Niedermeier, W., and Schwartz, M. C., Sodium and potassium in foods and waters: Determination by flame photometer. *J. Am. Dietet. Assoc.* **25,** 304–314 (1949).

SULFOBROMOPHTHALEIN (BSP) IN SERUM[*]

Submitted by: David Seligson and Jean Marino, Division of Biochemistry,
Graduate Hospital of the University of Pennsylvania, Philadelphia,
Pennsylvania

Checked by: John Reinhold, Pepper Laboratory of Clinical Medicine, Hospital
of the University of Pennsylvania, Philadelphia, Pennsylvania
Merle Lewis, Clinical Laboratory of Hammack and Maner, Los
Angeles, California

Introduction

Since the introduction of sulfobromophthalein (BSP) as a test
for liver function (2) it has gained wide acceptance and is considered
(3) the most reliable single test for evaluating liver function. BSP
extraction by the liver has also been used as a measure of hepatic
blood flow (4).

Principles

A measured amount of BSP, usually 5 (or 2) mg. per kilogram
of body weight, is injected intravenously. A normally functioning
liver removes the dye from the plasma proteins and excretes it into
the bile. If the liver function is impaired or biliary obstruction is
present, the excretion is delayed and a larger fraction of the dye
remains in the circulating blood 45 (or 30) minutes after the injec-
tion than remains when the liver is functioning normally. This
fraction is measured and expressed as amount of dye retention.

The concentration of the dye in serum or plasma is measured
colorimetrically after converting it from its colorless state to the
colored anion by the addition of alkali. The difference in color
between neutral and alkalinized serum is related to the concentration of BSP. In this method corrections for hemolysis, jaundice,
and turbidity are not required because these factors do not interfere
when the pH changes are controlled. Gaebler (5) and Reinhold (6)
apply correction factors for interferences.

[*] Based on the method of Seligson, Marino, and Dodson (1).

Binding of BSP to albumin with reduction of the dye extinction and effect on the spectral-absorbance curve is prevented by addition of the anion, p-toluenesulfonate, in large amounts.

Reagents

1. *Alkaline buffer, pH 10.6–10.7.* Dissolve 12.2 g. of Na_2HPO_4. $7H_2O$, 1.77 g. $Na_3PO_4.12H_2O$, and 3.20 g. sodium p-toluenesulfonate and make to 500 ml. with water. Adjust pH to 10.6–10.7 if necessary.

2. *Acid reagent, 2 M NaH_2PO_4.* Dissolve 69.0 g. of NaH_2PO_4. $1H_2O$ in water and make to 250 ml.

3. *BSP standard 10.0 mg./100 ml:* This solution is equivalent to the 100% retention standard for the 5 mg. per kilo test dose of BSP (7). Pure BSP[1] as "Bromsulphalein" can be obtained from Hynson, Westcott, and Dunning in Baltimore. Their test solution of 50.0 mg./ml. is diluted 1:500 with water to make the 10.0 mg./ 100 ml. standard. These solutions remain stable for 1 week, but deteriorate if held longer at room temperature.

Procedure

Place 0.50 ml. of serum into a test-tube cuvette and add 3.5 ml. of alkaline buffer. Mix gently and read in a photometer at or close to 580 mμ, using water as a reference blank. Add 0.10 ml. of acid reagent. Mix gently and read again. Proceed to the next sample.

Microanalyses can be made by placing 0.100 ml. of serum and 0.70 ml. alkaline buffer in a 1-cm. microcuvette.[2] Read the absorbance at 580 mμ in a Beckman Model DU or Model B spectrophotometer. Add 0.020 ml. acid reagent and read again.

Standard Curve

Measure the absorbance of dye concentrations equivalent to 50, 25, 10, 5, and 2.5% retention as described for samples and plot absorbance against the per cent retention. The standards are made by diluting the 10.0 mg./100 ml. standard to 5.0, 2.5, 1.0, 0.5, and

[1] Personal communication from Dr. John Brewer indicates that "Bromsulphalein" test solution varies from 98 to 102% of stated value and is, therefore, suitable for standardization of the sulfobromophthalein method.

[2] Pyrocell Manufacturing Company, 2075 East 84th Street, New York 28, New York; 50 x 3 x 10 mm. light path and a capacity of approximately 0.8 cc.

0.25 mg./100 ml. with water. These are respectively equivalent to 100, 50, 25, 10, and 5% retention when 5 mg. of dye is injected per kilogram of body weight. Since the curve follows Beer's law almost exactly, calculations may be based on the 5.0 mg./100 ml. or 50% retention standard:

$$\text{milligrams BSP/100 ml. serum} = \frac{\text{mg./100 ml. standard}}{\text{Absorbance of standard}}$$

$$\times \text{ absorbance difference (of serum)}$$

Per cent retention for 5 mg./kilo dose

$$= \frac{\text{Per cent retention standard}}{\text{Absorbance of standard}} \times \text{ absorbance difference (of serum)}$$

If the dose is 2 mg./kilo, the retention can be calculated by adjustments of standards. If the rate of disappearance of dye and per cent retention is desired, it can be calculated by the formula of Nadeau (8):

$$K = \frac{\log R_1 - \log R_2}{t_2 - t_1}$$

where K is the velocity constant, R_1 is the per cent retention at time 1 (t_1) in minutes, and R_2 is the per cent retention at time 2 (t_2) in minutes.

Discussion

In determining the dissociation curve of BSP, it was found that the pK for the dye is 8.8 and that 98% of the color is obtained between pH 7.4 and 10.4. Consequently, in the analysis these pH limits are used in order to attain a maximum color with minimum pH change.

In studying a series of absorbance measurements made of serums which were clear, hemolyzed, jaundiced, or turbid, it was found that greater absorbance changes occur in methods (7–9) where pH changes are uncontrolled than in the method described.

It was observed that the hue of BSP at pH 10.4 was purple but became more pink when albumin was present.

A spectral-transmittance curve of BSP at pH 10.3 in the presence of serum had an absorption maximum at 594 mμ. In the absence

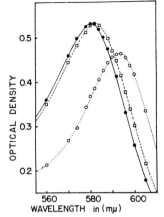

FIG. 1. Spectral-absorbance curve of sulfobromophthalein at pH 10.3 with and without albumin. Symbols: ●-BSP; □-BSP + albumin + PTSA; ○-BSP + albumin.

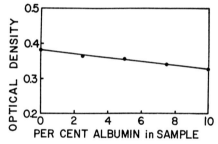

FIG. 2. Effect of albumin on optical density of sulfobromophthalein in serum. The abscissa represents grams per cent of an albumin solution containing BSP (25% retention) which was run as a hypothetical serum (0.5 ml. albumin solution plus 3.5 alkaline buffer).

of serum it was 580 mμ (Fig. 1). Furthermore, in the presence of serum there was less absorbance by the dye. Albumin solutions, simulating serum, without p-toluenesulfonate also gave lower absorbances. As the amount of albumin was increased, the absorbance of the dye decreased (Fig. 2). p-Toluenesulfonate was added in order to avoid this protein-binding effect. The amount chosen is approximately 29 μM/ml. in the final reaction mixture, which is 400 times the molar concentration of the albumin and 900 times that of BSP. With the addition of p-toluenesulfonate, the serum effect disappears and the absorption maximum is at 580 mμ.

Although the BSP dye is a relatively simple one, its measurement in the presence of albumin, bilirubin, hemoglobin, and turbidity can be difficult. The effects due to albumin have been overcome by the addition of p-toluenesulfonate as discussed above. The turbidity, usually due to lipoproteins and lipids, may be partly cleared by the high pH which occurs when NaOH is added for alkalinization (7–9). However, when the sample is acidified the turbidity may recur and in some cases was found to be more than was originally present in the serum. The problem of turbidity is made worse by delaying readings between alkalinization and acidification.

The hemoglobin presents a similar problem. The hemoglobin in serum, resulting from hemolysis, is converted in alkali to alkaline hematin which gives measurable absorbances at 580 mμ and depends on the time between formation and reading. Controlling pH with buffers as described prevents changes which might falsely increase the BSP content. Bilirubin also changes absorbance between high and low pH's. Jaundiced serum gives absorbance shifts with marked changes of pH.

In this method the pH shift is between 10.4 and 7.1. Over this range there is little or no absorbance change due to turbidity, jaundice, or hemolysis. Because both readings are made at one wavelength and no other corrections for interfering substances are required, speed is thereby achieved. Specificity appears to be increased by the controlled pH change.

Values in Healthy Persons

Serum from healthy adults is free from dyelike substances by this method. Healthy adults show less than 5% dye-retention 45 minutes after 5 mg. BSP per kilogram of body weight is injected. The 8–10% retention values for healthy adults stated in the literature probably reflect the inclusion of artifacts rather than impaired liver function.

REFERENCES

1. Seligson, D., Marino, J., and Dodson, E., The determination of Sulfobromophthalein in serum. *Clin. Chem.* **3**, 638–645 (1957).
2. Macdonald, D., Some observations on the disappearance of bromsulphalein dye from the blood: It's relation to liver function. *Can. Med. Assoc. J.* **39**, 556–560 (1938).
3. Zieve, L., and Hill. E., An evaluation of the factors influencing the discrimi-

native effectiveness of a group of liver function tests. III. Relative effectiveness of hepatic tests in cirrhosis. *Gastroenterology* **28,** 785–802 (1955).
4. Bradley, S. E., Ingelfinger, F. J., Bradley, G. P., and Curry, J. J., The estimation of hepatic blood flow in man. *J. Clin. Invest.* **24,** 890–897 (1945).
5. Gaebler, O. H., Determination of bromsulfalein in normal, turbid, hemolyzed or icteric serums. *Am. J. Clin. Pathol.* **15,** 452–455 (1945).
6. Reinhold, J. G., Bromsulfalein Tests. *In* "Medical and Public Health Laboratory Methods" (J. S. Simmons and C. J. Gentzkow, eds.), pp. 100–102. Lea and Febiger, Philadelphia, 1955.
7. Mateer, J. G., Baltz, J. I., Marion, D. F., and MacMillan, J. M., Liver function tests. *J. Am. Med. Assoc.* **121,** 723–728 (1943).
8. Nadeau, G., A simplified bromsulfalein clearance test. *Am. J. Clin. Pathol.* **24,** 740–746 (1954).
9. Rosenthal, S. M., and White, E. C., Clinical applications of bromsulphalein test for hepatic function. *J. Am. Med. Assoc.* **84,** 1112–1114 (1925).

UROBILINOGEN IN URINE AND FECES*

Submitted by: Bernard Balikov, Department of Metabolism, Walter Reed
Army Medical Center, Washington, D. C.
Checked by: Samuel Schwartz, University of Minnesota Medical School, Minneapolis, Minnesota
Marjorie Knowlton, Walter Reed Institute of Research, Walter
Reed Army Medical Center, Washington, D. C.
John G. Reinhold and Corinne Goldberg, Pepper Laboratory of
Clinical Medicine, Hospital of the University of Pennsylvania,
Philadelphia, Pennsylvania.
Ralph Jones, Baptist Memorial Hospital, San Antonio, Texas.

Introduction

The methods herein reported are based on the reaction of urobilinogen with p-dimethylaminobenzaldehyde (Ehrlich's aldehyde reagent) to form a red color. In 1931, an excellent review by Watson (1) mentioned the following facts directly pertinent to the development of these methods. Urobilin was first described by Jaffe in 1868, and the compound obtained by the reduction of urobilin was given the name urobilinogen by Le Nobel in 1887. It has been known that Ehrlich's aldehyde reaction frequently occurred in pathologic urines, and in 1903, Neubauer showed this to be due to the presence of urobilinogen. Subsequently, Charnas employed this reaction in a spectrophotometric method for the examination of urine and later of feces. This method was considerably improved by Terwen when, in 1925, he used alkaline ferrous hydroxide to reduce urobilin to urobilinogen, which enabled him to assay both as a single substance. In addition, he observed that the color formed by indole and skatole with the Ehrlich reagent was stable only in the presence of a strong mineral acid. By adding sodium acetate to the Ehrlich reagent, he converted the hydrochloric acid to acetic acid, thus eliminating interference from these substances. The urobilinogen-aldehyde color is unaffected by this procedure.

* Based on the method of Schwartz, Sborov, and Watson (3).

As a result of a study of some of the details of Terwen's method, Watson (1) devised a more efficient and simpler procedure. He also considerably improved the accuracy of the determination when he obtained a crystalline mesobilirubinogen for use as a primary standard. In 1936, Watson (2) reported the substitution of petroleum ether (petroleum benzin) for diethyl ether, which eliminated the oxidation of urobilinogen frequently experienced with the latter. Furthermore, he pointed out the advantages of petroleum ether-sodium carbonate preservation of the sample. Schwartz et al. (3), in 1944, determined the optimal ratio of sodium acetate to Ehrlich's reagent and in addition adapted the visual colorimetric procedure for use with a photoelectric colorimeter. In 1947, Watson and Hawkinson (4) reported the use of a pontacyl dye mixture as a secondary standard. The importance of controlling acidification prior to petroleum ether extraction was reported by Balikov (5) in 1955.

I. UROBILINOGEN IN URINE

Urobilinogen may be determined on a 24-hour specimen subjected to ferrous hydroxide reduction and petroleum ether extraction or on a 2-hour specimen by the "rapid" method. In the former method there is essentially no interference from other chromogens and results are reported in terms of urobilinogen content. In the latter method, other chromogens which may be present interfere, so that the report is given in terms of "Ehrlich units," which the Minnesota workers used to define chromogenic material which gave the Ehrlich reaction. One Ehrlich unit gives the same amount of color as 1 mg. of urobilinogen. The rapid method has the advantages of being simpler and of requiring a minimum of cooperation from the patient. It has the disadvantage of not being as reliable as the reduction-extraction method.

A. REDUCTION-EXTRACTION METHOD

The reduction-extraction method is taken from the report by Schwartz et al. (3) as modified by Balikov (5) and uses the pontacyl dye for standardization according to Watson and Hawkinson (4).

Reagents

1. *Ferrous sulfate* ($FeSO_4$), *20%.* This solution is stable for 24 hours.

2. Sodium hydroxide (NaOH), 10% (2.5 N).

3. Ehrlich's reagent, modified. Transfer 0.7 g. of *p*-dimethylamino-benzaldehyde and 150 ml. of concentrated hydrochloric acid to a flask and swirl to dissolve. Add 100 ml. of water and mix.

4. Sodium acetate ($NaC_2H_3O_2$ $3H_2O$), reagent grade, saturated. To insure saturation, some undissolved solid should be present.

NOTE: In the hands of the submitter, no difference in results was found when the less expensive triple-hydrated salt was substituted for the recommended anhydrous salt. (13)

5. Glacial acetic acid, C.P. or A.R.

6. Petroleum ether, ACS, b.p. 30–60°C.

7. Stock standard solution. Weigh 0.1000 g. of pontacyl carmine 2B† into a 500-ml. volumetric flask. Add 0.5% acetic acid to dissolve and then to bring up to volume. Measure 25.00 ml. of this solution into a 1-l. volumetric flask. Add 0.0950 g. of pontacyl violet 6R, 150%† to this flask and make up to volume with 0.5% acetic acid.

7. Working standard solution. Transfer 20.40 ml. of this stock solution to a 100-ml. volumetric flask. Dilute to volume with 0.5% acetic acid.

Collection of Specimen

A 24-hour urine specimen is collected in a brown bottle containing 100 ml. of petroleum ether and 5 g. of anhydrous sodium carbonate. The sample is kept refrigerated during the collection period.

Procedure

Measure the volume of the sample but exclude the volume of petroleum ether.

Conversion to urobilinogen: Deliver 50.0 ml. of urine into an Erlenmeyer flask. Add 25.0 ml. of 20% $FeSO_4$ and 25 ml. of 10% NaOH. Mix thoroughly and let stand in the dark for 1 hour.

Purification: Centrifuge. In order to select a suitable volume for analysis, add 2.0 ml. of supernatant to 2.0 ml. of modified Ehrlich's reagent. Mix and add 4.0 ml. of saturated sodium acetate. If the color is intense, proceed with 5.00 ml. of superantant; if pale red,

† This can be obtained from E. I. DuPont de Nemours & Co., Inc., Wilmington, Delaware, or Hartman-Leddon Co., Philadelphia 39, Pennsylvania.

proceed with 10.0 ml. Transfer the volume decided upon to a separatory funnel. Add water to bring the volume to 49 ml. Add 1.0 ml. of C.P. glacial acetic acid and mix. Add 35 ml. of petroleum ether and shake vigorously for about 7 minutes. After separation of the layers, discard the lower aqueous layer. Wash the petroleum ether twice with 25 ml. of water, discarding the water after each washing.

Color development: Add 3.0 ml. of Ehrlich's reagent and shake vigorously. Add 9.0 ml. of saturated sodium acetate. Shake. (Start a blank by adding 35 ml. of petroleum ether, 3.0 ml. of Ehrlich's reagent and 9.0 ml. of saturated sodium acetate to a small separatory funnel. Shake.) After separation into layers, transfer the lower colored aqueous solutions to 100-ml. mixing cylinders. Repeat the extraction with Ehrlich's reagent and sodium acetate, collecting the aqueous layers in the mixing cylinders, until no further color is extracted. Mix and measure the volume of the total aqueous extract.

Reading: Read at 565 mμ, setting the blank at 100% transmittance (T). Determine the urobilinogen equivalent from the standard curve.

Calculation

Total volume of urine in milliliters $\times \dfrac{2}{(\text{milliliters of supernatant used})}$

$\times$ milliliters of colored extracts $\times \dfrac{\text{mg./100 ml. (from standard curve)}}{100}$

$= $ milligrams of urobilinogen per 24 hours.

Standardization

Measure volumes of the working standard solution and 0.5% acetic acid into a series of test tubes as indicated in Table I.

Transfer these solutions to cuvettes. (In the submitter's laboratory, 19 x 150-mm. cuvettes are used.) Read on a spectrophotometer or photoelectric colorimeter at 565 mμ, using the blank for the zero setting. The concentration of the unknown colored extract is determined from a standard curve prepared by plotting absorbance or per cent transmittance against the standard solutions on suitable graph paper.

TABLE I

STANDARDIZATION SOLUTIONS

Working standard solution (ml.)	Acetic acid, 0.5% (ml.)	Urobilinogen equivalent (mg./100 ml.)
0.00	20.00	Blank
0.84	19.16	0.025
1.67	18.33	0.051
3.34	16.66	0.10
5.00	15.00	0.15
6.67	13.33	0.20
8.34	11.66	0.25
10.00	10.00	0.30
13.34	6.66	0.40
16.66	3.34	0.50
20.00	0.00	0.60

Range of Values in Healthy Adults

In a report which was recently published (13), this submitter found a range of from 0.05 to 2.5 mg. per 24 hours for 95% of healthy adults. This range is based on 46 urine values—24 on males and 22 on females. No significant difference was found between these two groups, and all values were thus included in calculating the normal range. The distribution of values is lognormal and the mean of 0.36 is skewed towards the lower limit.

Discussion

Light and oxygen oxidize urobilinogen to urobilin, which does not produce a color with Ehrlich's reagent. The use of petroleum ether and of a brown collection bottle inhibit this oxidation. The sodium carbonate prevents the extraction of urobilinogen by the petroleum ether which would occur in an acid medium and also stabilizes the urobilinogen (2).

The use of ferrous ion in alkaline solution is to reduce any urobilin present back to urobilinogen before the reaction with Ehrlich's reagent.

Extraction with petroleum ether will be incomplete if the concentration of acetic acid is other than as specified. A single extraction for 7 minutes has been adequate in the hands of the submitter, as checked by negative recoveries from a second extraction. However, this should be checked by each individual performing the analysis.

In the laboratory of the submitter, the final urobilinogen-aldehyde color was found to be stable for at least 70 minutes both in the dark and in ordinary artificial light.

It is of utmost importance to complete an extraction with the modified Ehrlich's reagent before the sodium acetate is added. If, for example, the two reagents are added simultaneously, results will be from 50 to 70% low.

Significance: (See reference 6.) Bacteria in the intestinal tract reduce much of the bilirubin entering the intestine to urobilinogen. A portion of the urobilinogen is reabsorbed by the large intestine and enters the blood stream through the portal vein. If the liver is unable completely to remove the urobilinogen, it passes into the general circulation and is excreted into the urine by the kidneys.

In obstructive jaundice, bile pigments reach the small intestine in decreased amounts. As a result, the urinary urobilinogen is low or absent. Thus, a negative urobilinogen accompanied by jaundice is indicative of obstruction to the flow of bile. Antibiotics, by their action on intestinal bacteria, may also cause urine urobilinogen accompanying jaundice to be low or absent.

Increased values are found in conditions such as infectious hepatitis in which the liver is damaged, or in diseases which cause intravascular hemolysis. In the latter case, the excessive urobilinogen formation and reabsorption may lead to increased urinary excretion.

B. RAPID METHOD

This method is based on the report by Watson *et al.* (7) using the pontacyl dye standard curves according to Watson and Hawkinson (4).

Reagents

The same Ehrlich reagent and sodium acetate prepared for the reduction-extraction method are used in this procedure.

Collection of Specimen

The bladder is emptied at 2:00 P.M. and the patient is then given a glass of water. The bladder is emptied at 4:00 P.M., and this complete specimen is used for analysis. The analysis must be run immediately.

Procedure

Measure the volume of sample.

Test for bilirubin. If positive, mix 2.0 ml. of 10% $CaCl_2$ with 8.0 ml. of urine and filter. Multiply the final result by 1.25 to correct for this dilution.

Turn on the spectrophotometer to warm it up. Set the wavelength at 565 mμ.

Measure 2.5 ml. of well-mixed sample into each of two small flasks. Label one B for blank and the other U for unknown. (If a number of specimens are to be assayed, complete each sample individually before proceeding with the next.) Add 5.0 ml. of sodium acetate to B. Mix well. Add 2.5 ml. of Ehrlich's reagent slowly, with constant shaking. Using this mixture as a blank, set the photometer at 100% T (565 mμ). Add 2.5 ml. of Ehrlich's reagent to U. Mix rapidly and well. Immediately add 5.0 ml. of sodium acetate, mix, and read.

Calculation

$$\frac{\text{Mg./100 ml. (from standard curve)} \times 4 \times \text{(milliliters of specimen)}}{100}$$

$$= \text{Ehrlich units per 2 hours}$$

Standardization

The standard curve prepared for the reduction-extraction procedure is used for this method.

Discussion

The result is reported in Ehrlich units rather than milligrams of urobilinogen because chromogens other than urobilinogen are normally present which interfere.

To prevent the oxidation of urobilinogen to urobilin, there must be no delay in running the analysis.

If bile pigments are present, the final color will be green rather than reddish; hence bile pigments should be removed (8).

Sulfonamides, procaine, 5-hydroxyindoleacetic acid, and other compounds react with Ehrlich's reagent and may interfere.

The normal range is from 0.1 to 1.2 Ehrlich units per 2 hours.

Significance: In addition to the interpretation given for the reduction-extraction method, it should be noted that this rapid method may give false positive results since chromogens other than urobilinogen interfere. In instances of doubt, a check should be made by the reduction-extraction method.

II. UROBILINOGEN IN FECES

For urobilinogen determination in feces, specimens should be collected for 4 or 5 days. Aliquots are subjected to ferrous hydroxide reduction and petroleum ether extraction. A random specimen may be reduced and the filtrate analyzed by the "rapid" method in which the petroleum ether extraction is omitted. The advantages and disadvantages of each method are the same as stated for urine above.

A. REDUCTION-EXTRACTION METHOD

The method reported here is taken from Schwartz *et al.* (3) as modified by Balikov (5), using the pontacyl dye standard curves according to Watson and Hawkinson (4).

Reagents

The same reagents are used as for urine urobilinogen by the reduction-extraction method.

Collection of Specimen

Record the time of the first defecation without collecting this specimen. For the following 4 or 5 days, collect all feces in weighed containers and refrigerate. Record the time and date when the last stool is passed.

Procedure

Determine the combined weight of feces. Transfer specimens as completely as possible to a Waring Blendor, adding a measured amount of water sufficient for easy mixing. For the weight of the blend, calculate milliliters of added water as grams and add this to the weight of the feces.

Conversion to urobilinogen: Transfer a 5.00-g. sample of the blend to a mortar. Use a total of 145 ml. of water as follows: Add 5–10 ml. of water to the sample and grind to a paste. Add 40–45 ml. of water and, after further grinding, let stand for a minute. Add 50 ml. of 20% $FeSO_4$ to a large flask. Add the supernatant fine suspension of feces from the mortar. Add more water to the remaining feces; grind and pour off the supernatant. Repeat, using the last of the 145 ml. of water to wash the mortar clean. Add 50 ml. of 10% NaOH slowly, with shaking. After mixing, let stand in the dark at room temperature for 1 to 3 hours or until the supernatant is relatively colorless. Filter through a paper equivalent to Whatman No. 2. If the filtrate is highly colored [$Fe_2(SO_4)_3$ is brown], measure 50.0 ml. into a flask, add 25 ml. of 20% $FeSO_4$ and 25 ml. of 10% NaOH, and mix well. Let stand again in the dark at room temperature for 1 to 2 hours. If the solution is still highly colored, transfer 50.0 ml. of this second filtrate and continue as described above. Filter if necessary.

Purification: To 2.0 ml. of filtrate add 2.0 ml. of Ehrlich's reagent and mix. Add 4.0 ml. of saturated sodium acetate. If the color is intense, proceed with 5.00 ml. of supernatant; if pale red, proceed with 10.0 ml. Transfer the volume decided upon to a separatory funnel. Add water to bring the volume to 49.5 ml. Add 0.5 ml. of C.P. glacial acetic acid and mix. Add 35 ml. of petroleum ether and shake vigorously for about 7 minutes. After separation, discard the lower aqueous layer. Wash the petroleum ether extracts with 100 ml. of water, discarding the lower aqueous layer.

The color is developed in the same manner as for urine urobilinogen by the reduction-extraction method, beginning with the addition of 3.0 ml. of Ehrlich's reagent.

Calculation

If the specimen weight is 250 g. or more:

$$\frac{24 \times \text{total grams of feces blend}}{\text{Hours collected} \times 5} \times 1^*$$

$$\times \frac{\text{mg./100 milliliters (from standard curve)} \times 250}{100 \times \text{milliliters of filtrate used}}$$

$\times$ milliliters of colored extracts $=$ milligrams of urobilinogen per 24 hours.

If the specimen weight is less than 250 g., it probably does not represent a complete 4-day collection and the following equation is used:

* This factor is 2 if a second incubation was needed and 4 if a third was needed.

$$\frac{\text{Milliliters of colored extracts}}{\text{Milliliters of filtrate used}} \times \text{mg./100 ml. (from standard curve)} \times 1^*$$

$$\times \frac{250}{5} = \text{Milligrams of urobilinogen per 100 g.}$$

Standardization

The standard curve prepared for urine urobilinogen is used for this method.

Range of Values in Healthy Adults

In a report which was recently published (13), this submitter found a normal range of from 57 to 200 mg. per 24 hours for adult males, and from 8 to 150 mg. for adult females. These ranges are based on 42 fecal values—21 on males and 21 on females. As for urine urobilinogen excretions, the distribution of values is lognormal. The mean for the male group is 101 mg. per 24 hours, and for the female group, 40 mg. The probability that this difference in the means is due to chance alone is less than 1 in 100.

Discussion

The sample and reagents used in the conversion to urobilinogen are half the amounts recommended by the original authors. Manipulations were thus made easier, and smaller amounts of reagent were used, with little apparent loss in accuracy or precision.

On standing, urobilinogen in feces is oxidized to urobilin, which does not react with Ehrlich's reagent. This urobilin is reduced to urobilinogen with ferrous ion before the reaction with Ehrlich's reagent.

Extraction with petroleum ether will be incomplete if the concentration of acetic acid is other than that specified. As for urine urobilinogen assays, the completeness of extraction with petroleum ether should be checked.

Significance: (See references 6 and 9.) The metabolism of hemoglobin, particularly in the reticulo-endothelial system, results in the formation of bile pigments. The liver removes these pigments from the blood and excretes them into the bile. The bilirubin in the bile entering the intestine is transformed, through the action of bacteria, into urobilinogen. Consequently, variations in the fecal

* This factor is 2 if a second incubation was needed and 4 if a third was needed.

urobilinogen as found in hepatic and biliary tract disease are determined largely by the degree of impairment of bilirubin excretion by the hepatic cells, the degree of obstruction to the flow of bile, and the severity of any associated hemolytic process. In the presence of liver damage, as in cirrhosis and hepatitis, the bile pigments, including urobilinogen, are not efficiently removed from the blood stream. There is thus a decreased excretion of pigments from the liver into the gastrointestinal tract which results in decreased fecal urobilinogen. In the absence of hepatic or biliary tract disease, the urobilinogen content of the feces is then an index of the degree of blood destruction. Thus, increases are found in hemolytic anemia, with or without jaundice, paroxysmal hemoglobinuria, etc., and decreases are found in anemias not accompanied by increased destruction of erythrocytes.

B. RAPID METHOD

This method is based on the report by Watson *et al.* (7), using the pontacyl dye standard curves according to Watson and Hawkinson (4).

Reagents

The same reagents are used as for urine urobilinogen by the reduction-extraction method, except that petroleum ether is not used.

Collection of Specimen

Any *fresh* specimen may be used. Record consistency as: fluid (if it can be poured), mushy (if it does not retain a cylindrical shape), or normal.

Procedure

Transfer a 10.0-g. sample of feces to a mortar. Use a total of 190 ml. of water as follows (use 90 ml. if feces are "clay-colored."): Add 10–20 ml. of water to the sample and grind into a paste. Add 80–90 ml. of water and, after further grinding, let stand for a minute. Add 100 ml. of 20% $FeSO_4$ to a large flask. Add the supernatant fine suspension of feces from the mortar. Add more water to the remaining feces, grind and pour off the supernatant. Repeat, using the last of the water to wash the mortar clean. Add 100 ml. of 10%

NaOH slowly, with shaking. Stopper, mix well, and let stand in the dark at room temperature for 1 to 3 hours. Filter through a paper equivalent to Whatman No. 2. Dilute 5.0 ml. of filtrate to 50 ml. with water. Turn on the spectrophotometer to warm it up. Pipet 2.5 ml. of diluted filtrate in each of 2 cuvettes, one marked B for blank and one marked U for unknown. If a number of specimens are to be assayed, complete each sample individually before proceeding with the next. Add 5.0 ml. of sodium acetate to B. Mix well. Add 2.5 ml. of Ehrlich's reagent slowly, with constant shaking. Set the wavelength at 565 mμ and then adjust the photometer to 100% transmittance with this blank. Add 2.5 ml. of Ehrlich's reagent to U and mix rapidly and well. Immediately add 5.0 ml. of sodium acetate, mix and read.

Calculation

Mg./100 ml. (from standard curve) $\times$ 1600 = Ehrlich units/100 g.
    If feces were clay-colored:
Mg./100 ml. (from standard curve) $\times$ 1200 = Ehrlich units/100 g.

Standardization

The standard curve used for urine urobilinogen is used for this method.

Discussion

The normal range is from 100 to 300 Ehrlich units per 100 g.

Maclagen (10) modified this method by using a smaller aliquot of sample, emulsifying it in a test tube, and doubling the concentration of ferrous hydroxide.

III. PORPHOBILINOGEN IN URINE

A simple procedure has been described by Watson and Schwartz (11) for the detection of porphobilinogen in urine. Porphobilinogen forms a red-colored compound when it reacts with the modified Ehrlich's reagent. It may be differentiated from urobilinogen and indole compounds, which react similarly, by extracting the reaction mixture with chloroform. The colors produced from the former remain in the aqueous layer, while those of the latter go into the chloroform layer.

Collection of Specimen

A fresh urine specimen is used. If the analysis is to be delayed, add 0.5 g. of sodium carbonate per 100 ml. of urine.

Procedure

In a large test tube mix 5.0 ml. of urine and 5.0 ml. of Ehrlich's reagent. Add 10.0 ml. of saturated sodium acetate and mix well. Add 5–10 ml. of chloroform and shake for several minutes. Allow the chloroform to settle. A red color in the supernatant is positive for porphobilinogen.

Porphobilinogen is colorless but upon standing in acid urine it polymerizes to form chains of pyrrole nuclei of various lengths called porphobilins (12). Porphobilins are brown, and a urine which darkens upon standing may be suspected of undergoing such a change. The sodium carbonate preservative alkalinizes the urine and inhibits this polymerization.

References

1. Watson, C. J., The average daily elimination of urobilinogen in health and in disease, with special reference to pernicious anemia. *A. M. A. Arch. Internal Med.* **47**, 698–726 (1931).
2. Watson, C. J., Studies of urobilinogen. I. An improved method for the quantitative estimation of urobilinogen in urine and feces. *Am. J. Clin. Pathol.* **6**, 458–475 (1936).
3. Schwartz, S., Sborov, V., and Watson, C. J., Studies of urobilinogen. IV. The quantitative determination of urobilinogen by means of the Evelyn photoelectric colorimeter. *Am. J. Clin. Pathol.* **14**, 598–604 (1944).
4. Watson, C. J., and Hawkinson, V., Studies of Urobilinogen. VI. Further experience with the simple quantitative Ehrlich reaction. Corrected calibration of the Evelyn colorimeter with a pontacyl dye mixture in terms of urobilinogen. *Am. J. Clin. Pathol.* **17**, 108–116 (1947).
5. Balikov, B., A note on quantitative urobilinogen determinations. *Clin. Chem.* **1**, 264–268 (1955).
6. Cantarow, A., and Trumper, M., "Clinical Biochemistry," 4th ed., pp. 434–439. Saunders, Philadelphia, Pennsylvania, 1949.
7. Watson, C. J., Schwartz, S., Sborov, V., and Bertie, E., Studies of urobilinogen. V. A simple method for the quantitative recording of the Ehrlich reaction as carried out with urine and feces. *Am. J. Clin. Pathol.* **14**, 605–615 (1944).
8. Ham, T. H., ed., "A Syllabus of Laboratory Examinations in Clinical Diagnosis," pp. 328–329. Harvard Univ. Press, Cambridge, Massachusetts, 1952.
9. Cantarow, A., and Trumper, M., "Clinical Biochemistry," 4th ed., p. 421. Saunders, Philadelphia, Pennsylvania, 1949.

10. Maclagen, N. F., cited *in* "Practical Clinical Biochemistry" by H. Varley, p. 237. Interscience, New York, 1954.
11. Watson, C. J., and Schwartz, S., A simple test for urinary porphobilinogen. *Proc. Soc. Exptl. Biol. Med.* **47**, 393–394 (1941).
12. Ham, T. H., ed., "A Syllabus of Laboratory Examinations in Clinical Diagnosis," pp. 289–298. Harvard Univ. Press, Cambridge, Massachusetts, 1952.
13. Balikov, B., Urobilinogen excretion in normal adults, results of assays with notes on methodology. *Clin. Chem.* **3**, 145–153 (1957).

AUTHOR INDEX

Numbers in parentheses are reference numbers. They are included to assist the reader to locate references in which the authors' names are not listed. Italic numbers indicate pages on which references are listed.

A

Abell, L. L., 26, 27(2), *33*
Abul-Fadl, M. A. M., 129, *131*
Ahrens, E. H., Jr., 47(10), *48*
Albritton, E. C., 183, *185*
Aldrich, R. A., 145(9, 10), *146*
Allen, W. M., 65, *68*
Amdur, M. O., 47(7), *48*
Andersen, A. C., 93(14), *99*
Archibald, R. M., 91(2), 95(15), *98, 99*
Armstrong, A. R., 128, *130*
Arrowsmith, W. R., 74(12), 75(12), *78*
Ashworth, J. N., 44(3), *47*
Astrup, P., 111(10), 119, *120*
Austin, J. H., 49, 53(12), 58(12), *60*

B

Baker, R. W. R., 165(1), *184*
Balikov, B., 193, 194(13), 196(13), 199, 201, *204, 205*
Baltz, J. I., 17(19), *21*, 187(7), 188(7), 190(7), *191*
Bangerter, F., 1(1), *11*
Bardawill, C. J., 42(2), *47*
Barkan, G., 69, 70, 76(4), *77*
Barker, S. B., 147, 162(20), *163*
Bashaur, F., 145, *146*
Bates, R. G., 108, 109(2), 112(2), 114(2), 115, 118(2), *120*
Beckman, W. W., 91(3), 95(15), *98, 99*
Belk, W. P., 56, *60*
Bell, R. D., 122, *129*
Berkman, S., 86(2), 87(2), 88(2), 89(2), *90*
Bernard, A., 2(11), *11*
Bernhart, F. W., 49(2), *59*
Bertie, E., 197(7), 202(7), *204*
Bessey, O. A., 128(17), *130*
Bethard, W. F., 76(21), *78*
Biedermann, W., 1(1), *11*

Bierman, H. R., 145(8), *146*
Bills, C. E., 183, *185*
Blackburn, C. M., 162(21), *163*
Blonstein, M., 47(9), *48*
Bloom, B., 39(5), *39*
Bockus, H. L., 86(8), 89(8), *90*
Bodansky, A., 122, *129, 130*
Bodansky, M., 129, *130, 131*
Bodansky, O., 129, *130, 131*
Bongiovanni, A. M., 61, 66(4), 67(7), *68*
Bonner, C. D., 129(32, 34), *131*
Bortolotti, T. R., 1(4), 6(4), *11*
Bossenmaier, I., 144(7), *146*
Boyle, A. J., 147(9), *163*, 165(4, 7, 12), *184*
Bradley, A. F., 111(11), 118(11), 120(17), *120, 121*
Bradley, G. P., 186(4), *191*
Bradley, S. E., 186(4), *191*
Bradstreet, R. B., 92(4, 5), 93(4, 5), *99*
Brand, F. C., 27, *33*
Brechbühler, T., *164*
Brendler, H., 128(18), *130*
Briggs, A. P., 122, *129*
Brock, M. J., 128(17), *130*
Brodie, B. B., 26, 27(2), *33*
Brown, D. M., 47(5), *48*
Brown, H., 162(18), *163*
Bruger, M., 12(4), *21*
Buckley, E. S., Jr., 1, 6, *11*
Bunch, L. D., 86(1), 87(1), 88(1), 89(1), *90*
Bunn, D., 13(12), *21*
Butler, A. M., 147(6), *163*

C

Callow, N. H., 79, 80, 84(2), *85*
Callow, R. K., 79, 80, 84(2), *85*
Calvary, E., 40, 45(1), 47(1), *47*

207

SUBJECT INDEX

A

Acid phosphatase, (see Phosphatase, acid)
Adrenal function, ACTH infusion test, 66
Alcohol, ethyl, purification, 62, 80
  for 17-hydroxycorticoids, 62
  for 17-ketosteroids, 80
Alkaline phosphatase, (see Phosphatase alkaline)
Aminonaphtholsulfonic acid, reagent, 124, 133
  purification, 124
Ammonium chloride, standard, 96
Ammonium purpurate, reagent, 3
Ammonium sulfate, standard, 101
Arsenite reagent, 151, 152

B

Bathophenanthroline, 77
B-glycerophosphate substrate, 124
Biuret, reaction, 47
  reagent, 41–42
  Gornall, 42
  Weichselbaum, 41
Bodansky unit, 123
Bromsulphalein (see Sulfobromophthalein)
BSP (see Sulfobromophthalein)
Buffers, standard, (see pH Standards)

C

Calcium, determination, compleximetric, in serum, 1–11
  spectrophotometric titration, 3
  standard for, 3
Cephalin, 132
Cephalin-cholesterol flocculation test, 12–21
Cephalin-cholesterol reagent, 14, 19

Cerebrospinal fluid
  chloride, determination in, 24
  total nitrogen, determination in, 98
Ceric-arsenite reaction, 147
  inhibition by Brucine, 147
  inhibition by Mercury, 147
  reagents for, 151, 152
Ceric sulfate reagent, for PBI, 151, 161–162
Chloride, determination, 22–25
  range of values, 24
  in serum, 23
  in spinal fluid, 24
  standard for, 23
  in urine, 25
Cholesterol, determination, 26–33
  comparison with Schoenheimer and Sperry, 32
  in serum or plasma, 27
  standard for, 27
Cortisone standard, 63
Chromic acid, for PBI, 151
Coproporphyrin, 137–146
  determination, 142
  purification, 139–140
  range of values, 145
  standard, 139, 140
  (see also Porphyrins in urine)

D

Dehydroepiandrosterone, standard, 80
Dichlorofluorescein indicator, 23
m-Dinitrobenzene, 80
  purification, 80
  reagent, 80

E

EDTA, reagent, 3
Ehrlich's reagent, 194
"Ehrlich unit" definition of, 193, 198
Electrodes, 108–112

213

Kjeldahl, nitrogen (see Nitrogen-Kjeldahl)

L

Lecithin, 132
Liebermann-Burchard reagent, 27
Lithium, internal standard, 173, 181
Lipase, determination in serum 86–90
(see Pancreatitis-lipase)
range of values, 89
unit, definition of, 89

M

Macro-Kjeldahl, 92 (see also Nitrogen, Kjeldahl)
Micro-Kjeldahl, 96, 100
for total nitrogen, 98
for NPN, 100
(see also Nitrogen, Kjeldahl)
Molybdate reagent, 123, 133
Murexide reagent, 3

N

Nessler's solution, 101
Nitrogen, Kjeldahl, determination, 91–99
cerebrospinal fluid, 98
compounds determinable by Kjeldahl, 91–92
factor for conversion to protein, 95
macro-Kjeldahl method, 92
in feces, 95
in serum, 92
in urine, 95
mercury catalyst, 93
micro-Kjeldahl method, 96
in cerebrospinal fluid, 96
in feces, 96
NPN in blood, 96
in urine 96
nitrates, 93, 95
nitrites, 93, 95
nitro derivative, 93
standard, 96
Non-protein nitrogen, determination, 100–106
constituents, 104
in blood or plasma, 102–104
protein-free filtrate, 102

range of values, 104
in urine, 95
NPN, 98, 99, 100
(see also Non-protein nitrogen, determination)

O

Olive oil substrate, for lipase determination, 87

P

Pancreatitis-lipase, 86–90
definition, 87
range of values, 89
substrate, 86–87
olive oil, 87
tributyrin, 86–87
PBI (see Protein-bound iodine)
pH, determination in blood, 107–121
collection of blood, 113–114
colorimetric, 112
definition, 107
electrodes, 108–111 (see also Electrodes)
errors, 117–118
detection of, 117–118
meters for, 108–109
micromethod, 116
open electrode, 116
range of values, 119
standard buffers, 115 (see also pH standards)
temperature correction, 110–111, 116
pH standards, 115
phosphate buffer, pH 7.38, 115
potassium hydrogen phthalate, pH 4.01, 115
sodium borate pH-9.18, 115
1, 10–Phenanthroline reagent, 71
Phenylhydrazine-sulfuric acid reagent, 62
Phosphatase, 122–131
range of values, 128
(see also Phosphatase, acid, Phosphatase, alkaline)
Phosphatase, acid, 122–131
determination in serum, 127
prostatic, 129
tartrate inhibition of, 129
range of values, 129